Stephan Kaufmann

A Crash Course
in *Mathematica*

Birkhäuser Verlag
Basel · Boston · Berlin

Author:

Stephan Kaufmann
Mechanik
ETH Zentrum
CH-8092 Zürich

E-mail: kaufmann@ifm.mavt.ethz.ch
Homepage: http://www.ifm.ethz.ch/~kaufmann

1991 Mathematics Subject Classification 00-01

A CIP catalogue record for this book is available from the Library of Congress, Washington D.C., USA

Deutsche Bibliothek Cataloging-in-Publication Data

A Crash Course in Mathematica [Medienkombination] / Stephan
Kaufmann. - Basel ; Boston ; Berlin : Birkhäuser
  ISBN 3-7643-6127-1

©1999 Birkhäuser Verlag, Postfach 133, CH-4010 Basel, Schweiz
Cover design: Markus Etterich, Basel
Printed on acid-free paper produced of chlorine-free pulp. TCF ∞
Printed in Germany
ISBN 3-7643-6127-1
ISBN 0-8176-6127-1

9 8 7 6 5 4 3 2 1

■ Contents

■ Preface

● About Mathematica

Mathematica unites the following tasks, among others, in one uniform interactive environment:
• the entry and display of mathematical formulas,
• numerical calculation,
• symbolic mathematics,
• plotting functions,
• contours and density plots
• parametric plots of curves and surfaces,
• creating graphics from elementary objects,
• animating graphics,
• list processing,
• pattern matching,
• functional, procedural and rule-based programming,
• structuring documents hierarchically,
• programming interactive documents.
This is the ideal tool for those who use pure or applied mathematics, graphics, or programming in their work.

Mathematica is available for all the usual computer operating systems. Thanks to the uniformity of its file format, it is also a practical medium for the electronic exchange of reports or publications which contain formulas and graphics. *Mathematica* files, called *notebooks*, can also be saved directly into HTML format for easy publication on the World Wide Web.

Mathematica allows you to solve many problems quickly, like calculating integrals, solving differential equations, or plotting functions. In order to use this powerful tool efficiently, however, you need to know the basics of the user interface and of the syntax of *Mathematica* expressions. Otherwise you would be like a driver who has not noticed that there are more gears than just first and that it makes sense to obey the rules of the road. In both cases its better not to attempt to learn by just trying things out.

- **The Goals of this Course**

This book and the accompanying *Mathematica* notebooks on CD-ROM give you the basics of *Mathematica* in short form. We will discuss the user interface (*front end*), the most important functions built into the actual calculator (*kernel*), and some additional programs (*packages*) which come with *Mathematica*. The examples are kept at a simple mathematical level and to a great extent independent of special technical or scientific applications. Emphasis is put on solving standard problems (equations, integrals, etc.) and on graphics.

After working through this course you will be able to solve your own problems independently and to find additional help in the online documentation. Depending on your interests and needs, completing the first two parts of this course may be sufficient, as they include the most important calculations and graphics functions. The third part is more technical and the fourth introduces programming with *Mathematica*.

- **The Book and the CD-ROM**

The book is basically a direct printout of the corresponding *Mathematica* notebooks on the CD-ROM. Some things had to be left out like the colors, the animation of graphics, and also the hyperlinks within the notebooks to the online documentation of *Mathematica* and to Web sites.

Why a book? Books are still the most ergonomic medium for the sequential study of texts–and today most of them are still lighter than a laptop computer.

- **What this Course Is Not**

This course is neither complete nor meant to be a reference tool. The four parts of the book therefore do not include summaries of the *Mathematica* commands discussed. However, the notebooks on the CD-ROM contain hyperlinks to online documentation of the commands. The advantage being that you always see the documentation corresponding to your version of the program.

A complete installation of the program includes the 1403 page "*Mathematica Book*" by Stephen Wolfram. This book is perhaps the first exception to the rule above: because of its size and format, comparable to a laptop, the electronic version, with its many useful hyperlinks, is usually more practical than the printed version.

- **Organization**

The introduction contains a short overview of *Mathematica*'s capabilities and–for minimalists–a summary of the most important commands. The following four parts form a progression and should therefore be done in sequence. It is not necessary, however, to complete all the parts in one go. The methods in the first two parts will already allow you to solve many problems. The motivation for studying the last two parts will probably arise after you have worked with the program for a while.

The *first part* leads to the most important capabilities of the user interface (*front end*) and explains the different possibilities for creating *Mathematica* entries and formulas. Next, how to tackle the most common problems is shown using examples: numerical calculation, manipulation of formulas, solving equations and differential equations, calculating limits, derivatives and integrals.

The *second part* deals with an especially compelling aspect of the program: plots of graphs of functions and parametric plots of curves and surfaces. Many of these features are built into the *Mathematica* kernel; additional useful tools are available from standard packages.

The *third part* starts with a discussion of lists. They are used to manipulate vectors and matrices; they also appear in many *Mathematica* functions as arguments or results, and can be used to structure data. In connection with this, this part also deals with mapping functions on lists and simple calculations of linear algebra. Lists allow you to assemble graphics from graphics elements (lines, circles, etc.). Sequences of graphics can be animated.

The *fourth part* is aimed at users who want a more in-depth study of *Mathematica*. It is the starting point for the independent development of complicated programs. The first three chapters are dedicated to the structure and evaluation of *Mathematica* expressions. Based on this, we discuss different possible programming methodologies and the tools for their application. At the end you will find leads to further information such as relevant Web sites and a link to *Mathematica* literature.

Several chapters include *in-depth paragraphs* covering special features and technical details, which can be left out at first.

The *exercise problems* have been kept simple on purpose. They should allow you to master the program without getting bogged down in complicated mathematics. The ideal exercise examples are not found in the book–they develop from your work. There are many problems which you can solve with *Mathematica*. Try it!

- **Tips**

For best results, the notebooks should be worked on directly in *Mathematica* on the computer. If you do not own the complete program, you can use the program *MathReader,* which is included on the CD-ROM, to access the notebooks (and the animations). *Math-Reader* is a reduced version of *Mathematica* which cannot be used to make calculations but which does give you a first impression of how the program works.

When using the full version, it is best to use the files in the `In-only` directory; for *MathReader* use the files in `In-out` (see the paragraph "The Files on the CD-ROM").

It is important to know that the cell groups (shown as square brackets on the right-hand side of the notebook window) can be opened or closed by double-clicking on the bracket itself, or by using the command **Cell > Cell Grouping** on the menu bar.

With the menu **Format > Magnification** you can adjust the magnification of the window for maximum overview and readability. Graphics might then appear jaggy. Use the command **Cell > Rerender Graphics** to smooth them out again.

With the computer you can use the hyperlinks to access the documentation of built-in functions, or to jump from one section of the book to another. The menu **Find > Go Back** is useful here: it takes you back to the original hyperlink. Depending on the version and the installation options of *Mathematica*, certain links are inactive. The links in the table of contents and the subject index are useful to navigate between the notebooks.

It is best to start with the examples in the chapter "A Short Tour" (in the `Introduction.nb` file). With the full version of *Mathematica* the input cells can be evaluated using the <Enter> key (or <Shift> and <Return>). In the "Short Tour", and during the whole course, you are invited to change the examples in order to test the possibilities and limits of the program and to get used to the syntax.

It will quickly become obvious that a lot can be done with the commands in the "Short Tour", but that much remains unclear. This will motivate you towards a systematic and in-depth study of the program using the rest of the course sections.

- **The Files on the CD-ROM**

The CD-ROM can be used with MacOS, Windows 95/98/NT, or UNIX. It contains the *Mathematica* notebooks from the book in different versions, as well as (for MacOS and Windows) the program *MathReader*, with which the notebooks and the animations can be viewed but not changed.

The file `Info.txt` contains up-to-date information.

The actual notebooks are named according to their contents:
- Contents.nb,
- Introduction.nb,
- Part-1.nb to Part-4.nb,
- Index.nb.

They are filed in two versions: with and without the *Mathematica* output cells. The files with the output cells (In-out folder) are much larger than those without (In-only folder), mainly because of the graphics.

If you work with the complete version of *Mathematica*, it is best to use the notebooks that contain only the input cells (In-only folder). You can evaluate them using the <Enter> key (or <Shift> and <Return>) and thus reproduce the full notebook.

The files in the In-out folder contain all the input and output cells. They are meant to be viewed with *MathReader*.

The second and third sections contain the most graphics. Depending on the magnification and the number of graphics and animations already viewed, *Mathematica* or *MathReader* will need a large amount of memory. It is therefore recommended to only have one notebook open at a time. If you are using a computer with static memory assignment (Macintosh) you should assign *Mathematica* or *MathReader* as much memory as possible. In doing so a compromise between the front end (*Mathematica*) and the kernel (*MathKernel*) must be found.

- **Information About this Book on the World Wide Web**

Up-to-date information and any corrections to the book and the files on the CD-ROM can be accessed on the Web at http://www.ifm.ethz.ch/~kaufmann/.

- **Technical Information**

The notebooks were created and evaluated with *Mathematica* 3.0.1 on a PowerMacintosh 8600/200. The beginning of each new kernel session can be identified by the numbers of the input cells (In[...]).

The Postscript files used to print the book were created directly from the notebooks using a test version of *Mathematica* 4.0 (which allows automatic hyphenation).

The format is based on the default *Style Sheet* (**Format** > **Style Sheet** > **Default**), with some additional header and body text styles.

The only difference to the default settings of the kernel is a new definition of $Default-Font, created to use a smaller font size in the graphics. The definition reads:

```
$DefaultFont = {"Courier", 9}
```

It was added to the init.m file in the Configuration/Kernel subdirectory of the *Mathematica* installation folder.

Using the **Option Inspector** (**Format** menu), the ImageSize for normal graphics was set at 250×250 points, and at 220×220 points for the smaller graphics in the exercise and in-depth sections. Further changes in ImageSize were added directly in each graphic command and can be deleted during your work with the notebooks.

In the notebooks in the In-out folder (see "The Files on the CD-ROM") the option CellLabelAutoDelete was set to False with the **Option Inspector**, so that the numbers of the input and output cells would remain after closing the notebooks.

The subject index was created with a test version of the *AuthorTools* package from Wolfram Research.

• Acknowledgements

Many people contributed to the success of this project and deserve my heartfelt thanks:

• Dr. Thomas Hintermann and the Birkhäuser Verlag for their spontaneous interest and efficient realization,

• my wife Brigitta for her love and strength in during the "blessed" year of 1998 and her proofreading of the German manuscript,

• Tobias Leutenegger and Frank May for their correction of many mistakes in the German manuscript,

• Mathias Götsch for his help in preparing the CD-ROM,

• Dianne Littwin, Jamie Peterson, and Andre Kuzniarek of Wolfram Research for their help with *MathReader*, *AuthorTools* and test versions of *Mathematica*,

• Prof. Mahir Sayir for his farsighted and liberal management of the Institute for Mechanics, which allows the motivation and the freedom for projects like this,

• Prof. Jürg Dual and the other "young professors" of the Department of Mechanical and Process Engineering at ETH through the launching of "Engineering Tools" courses, one of which, the "Software for Symbolic Mathematics", I gave, which in turn spawned these notebooks,

• Prof. Urs Stammbach for his valuable suggestions and his in-depth group, from which I was able to recruit students to look after the course,

• the second-semester students of mechanical and process engineering at ETH, who took an active part in the course in spring 1998.

• About this English Translation

This is basically a direct translation of the German original "*Mathematica* – kurz und bündig" (Birkhäuser, 1998). Only a couple of details have been changed or added to clarify certain points. The author is very grateful to Katrin Gygax for her excellent translation.

■ A Short Tour

This section introduces the most important features of *Mathematica*, using simple examples.

■ Formula Entry

Formulas can be entered using various techniques with palettes or using only the keyboard.

■ Entries Using Palettes

The menu **File > Palettes > BasicInput** displays a palette with the simplest formulas on-screen. You can use this to create an exponent, for example.

$$\square^{\square}$$

Now enter 2.

$$2^{\square}$$

Use the tab key to jump to the next placeholder and enter 3.

$$2^3$$

Pressing the <Enter> key (or <Shift> and <Return>: $\boxed{\text{SHIFT}}\boxed{\text{RET}}$) evaluates the cell.

In[1]:= 2^3

Out[1]= 8

■ Entries Using the Keyboard

The exponent can also be written using ^. This gives us the equivalent keyboard entry.

In[2]:= $2 \wedge 3$

Out[2]= 8

Even the "two-dimensional" 2^3 can be done on the keyboard: enter 2 $\boxed{\text{CTRL}}^\wedge$ 3.

In[3]:= 2^3

Out[3]= 8

■ Numerical Calculations

Mathematica is not only a formula editor but also an expensive pocket calculator that can still do a thing or two.

■ Exact Arithmetic

We can calculate with exact integers and rational numbers of any size.

In[4]:= **2⁵¹²**

Out[4]= 1340780792994259709957402499820584612747936582059239337772357¬
         6144372176403007354697680187429816690342769003185818648605¬
         085375388281194656994643364900608409

In[5]:= **2 ^ 10 / 10 ^ 3**

Out[5]= $\frac{128}{125}$

■ Arithmetic with Approximate Numbers

Numerical approximations of varying precision are possible.

In[6]:= **N[π, 200]**

Out[6]= 3.1415926535897932384626433832795028841971693993751058209749¬
         4459230781640628620899862803482534211706798214808651328230¬
         6647093844609550582231725359408128481117450284102701938521¬
         105559644622949489549303820

■ Arithmetic with Complex Numbers

Complex numbers are entered using the imaginary unit I (or *i*).

In[7]:= **(1 + 3 I) ^ 2**

Out[7]= -8 + 6 I

■ Symbolic Mathematics

By using symbol names instead of numbers we get mathematical expressions. These can be manipulated, just like calculations "by hand".

■ Polynomials

This is a polynomial in three variables:

In[8]:= **(a + b + c) ^ 5**

Out[8]= $(a + b + c)^5$

The Expand function expands it out.

In[9]:= **Expand[(a + b + c) ^ 5]**

Out[9]= $a^5 + 5 a^4 b + 10 a^3 b^2 + 10 a^2 b^3 + 5 a b^4 + b^5 + 5 a^4 c + 20 a^3 b c + 30 a^2 b^2 c + 20 a b^3 c + 5 b^4 c + 10 a^3 c^2 + 30 a^2 b c^2 + 30 a b^2 c^2 + 10 b^3 c^2 + 10 a^2 c^3 + 20 a b c^3 + 10 b^2 c^3 + 5 a c^4 + 5 b c^4 + c^5$

■ Equations

We solve the equation $x^3 + x^2 - x + 1 = 0$ for x like this:

In[10]:= **Solve[x^3 + x^2 - x + 1 == 0, x]**

Out[10]= $\left\{\left\{x \to -\dfrac{1}{3} - \dfrac{4}{3\,(19 - 3\sqrt{33}\,)^{1/3}} - \dfrac{1}{3}\,(19 - 3\sqrt{33}\,)^{1/3}\right\},\right.$

$\left\{x \to -\dfrac{1}{3} + \dfrac{2\,(1 + I\sqrt{3}\,)}{3\,(19 - 3\sqrt{33}\,)^{1/3}} + \dfrac{1}{6}\,(1 - I\sqrt{3}\,)\,(19 - 3\sqrt{33}\,)^{1/3}\right\},$

$\left.\left\{x \to -\dfrac{1}{3} + \dfrac{2\,(1 - I\sqrt{3}\,)}{3\,(19 - 3\sqrt{33}\,)^{1/3}} + \dfrac{1}{6}\,(1 + I\sqrt{3}\,)\,(19 - 3\sqrt{33}\,)^{1/3}\right\}\right\}$

The function FindRoot returns an approximate solution of a transcendental equation.

In[11]:= **FindRoot[Sin[x] + 1 == x, {x, 2}]**

Out[11]= $\{x \to 1.93456\}$

■ Derivatives

The following expression calculates the derivative of $x^{\sin(x^{\cos(x)})}$ for x.

In[12]:= **D[x^Sin[x^Cos[x]], x]**

Out[12]= $x^{\mathrm{Sin}[x^{\mathrm{Cos}[x]}]}\,\mathrm{Cos}[x^{\mathrm{Cos}[x]}]\,\mathrm{Log}[x]$
$(x^{-1+\mathrm{Cos}[x]}\,\mathrm{Cos}[x] - x^{\mathrm{Cos}[x]}\,\mathrm{Log}[x]\,\mathrm{Sin}[x]) + x^{-1+\mathrm{Sin}[x^{\mathrm{Cos}[x]}]}\,\mathrm{Sin}[x^{\mathrm{Cos}[x]}]$

▪ Integrals

Using the template from the palette **BasicInput**, we create an integral and calculate it.

In[13]:= $\int \dfrac{1 + x^2 + x^3 - 3\,x^5}{(1 + x - x^2)^2}\, dx$

Out[13]= $-6\,x - \dfrac{3\,x^2}{2} + \dfrac{6\,(3 + 4\,x)}{5\,(-1 - x + x^2)} +$

$\dfrac{152\,\mathrm{ArcTanh}\left[\frac{-1 + 2\,x}{\sqrt{5}}\right]}{5\,\sqrt{5}} - 7\,\mathrm{Log}\left[-1 - x + x^2\right]$

This is another way of writing integrals:

In[14]:= **Integrate[Sin[x]^3 Cos[x]^2 Exp[x], x]**

Out[14]= $\dfrac{1}{2080}\,(\mathrm{E}^x\,(-130\,\mathrm{Cos}[x] - 39\,\mathrm{Cos}[3\,x] +$

$25\,\mathrm{Cos}[5\,x] + 130\,\mathrm{Sin}[x] + 13\,\mathrm{Sin}[3\,x] - 5\,\mathrm{Sin}[5\,x]))$

▪ Graphics

Various graphics functions can be used to visualize mathematical functions and mappings (or data) in all ways imaginable.

▪ Two-Dimensional Graphics

First we plot the graph of the function $x \to \dfrac{x^2 - x + 3}{x^3 - 2\,x^2 - 1}$ in the interval $[-10, 10]$.

In[15]:= **Plot$\left[\dfrac{x^2 - x + 3}{x^3 - 2\,x^2 - 1}, \{x, -10, 10\}\right]$**

Out[15]= ▪ Graphics ▪

This parametric plot creates a spiral:

In[16]:= **ParametricPlot[{φ Cos[φ], φ Sin[φ]}, {φ, 0, 2 π}]**

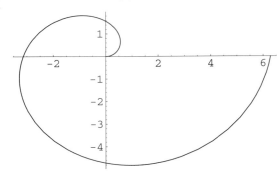

Out[16]= - Graphics -

■ Three-Dimensional Graphics

The following command plots the graph of the function $(x, y) \rightarrow \sin(x\,y)$.

In[17]:= **Plot3D[Sin[x * y], {x, 0, 2 * Pi}, {y, 0, 2 * Pi}]**

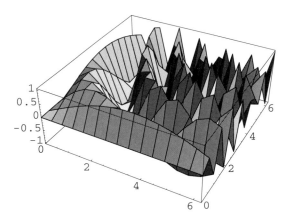

Out[17]= - SurfaceGraphics -

The peaks can be smoothed out by increasing the number of function values calculated initially. We also use a more elegant way of writing the input:

In[18]:= `Plot3D[Sin[x y], {x, 0, 2 π}, {y, 0, 2 π}, PlotPoints → 40]`

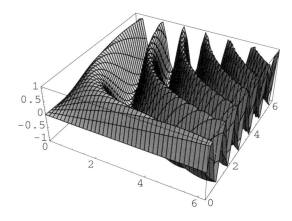

Out[18]= - SurfaceGraphics -

Functions of two variables can also be visualized using contours.

In[19]:= `ContourPlot[x² - y², {x, -2, 2}, {y, -2, 2}, PlotPoints → 30]`

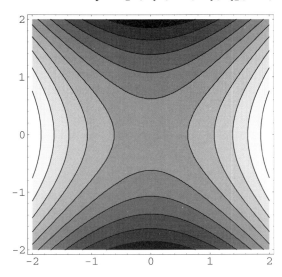

Out[19]= - ContourGraphics -

■ Animated Graphics

Sequences of graphics can be animated on-screen. This expression creates a graphic sequence of two colored lines:

In[20]:= `Table[Show[Graphics[{Thickness[0.05], {Hue[`$\frac{t}{\pi}$`], Line[`

`{-{Cos[t], Sin[t]}, {Cos[t], Sin[t]}}]}, {Hue[`$\frac{t}{\pi} + \frac{1}{2}$`],`

`Line[{{-Sin[t], Cos[t]}, {Sin[t], -Cos[t]}}]}}],`

`PlotRange → {{-1, 1}, {-1, 1}}, AspectRatio → Automatic,`

`ImageSize → 150], {t, 0,` $\frac{\pi}{2} - \frac{\pi}{30}$`,` $\frac{\pi}{30}$`}];`

Double-clicking the graphic rotates the cross. The book shows only the first position. Therefore another representation of all 15 positions is given below:

In[21]:= `Show[GraphicsArray[Partition[%, 5]]]`

Out[21]= - GraphicsArray -

■ Programming

Mathematica is a powerful high-level programming language that supports functional and rule-based programming as well as the usual procedural programming styles.

As an example let us look at a program for the recursive calculation of factorials. All we need are the following two definitions:

```
In[22]:=  fac[0] = 1;
          fac[n_] := n fac[n - 1]
```

The result for 100! yields

```
In[24]:=  fac[100]
```

```
Out[24]=  9332621544394415268169923885626670049071596826438162146859296
          3895217599993229915608941463976156518286253697920827223758%
          251185210916864000000000000000000000000
```

and matches the one produced by the built-in factorial function:

```
In[25]:=  100 !
```

```
Out[25]=  9332621544394415268169923885626670049071596826438162146859296
          3895217599993229915608941463976156518286253697920827223758%
          251185210916864000000000000000000000000
```

■ Some of the Most Important Functions

This short overview only gives a quick description of 33 important *Mathematica* functions. The selection must be arbitrary because there are more than 1600 objects built into the kernel of Version 3.0. The online documentation in *Mathematica* (see Section 1.2) contains more precise and up-to-date information on all built-in functions. In the notebook you can just click on the hyperlinks to get there.

■ Numerical Approximations

N[x]	numerical approximation of an expression
N[x, n]	numerical approximation with n digits

■ Constants

Pi	$\pi \approx 3.14159$
E	$e \approx 2.71828$
I	$i = \sqrt{-1}$

■ Elementary Functions

`Sqrt[x]`	square root
`Exp[x]`, `Log[x]`	exponential function, natural logarithm
`Sin[x]`, `Cos[x]`, `Tan[x]`	trigonometric functions
`Sinh[x]`, ...	hyperbolic functions
`ArcSin[x]`, ...	inverse trigonometric functions
`ArcSinh[x]`, ...	inverse hyperbolic functions

■ Manipulation of Expressions

`Expand[x]`	expand out
`Factor[x]`	factor
`Simplify[x]`, `FullSimplify[x]`	simplify

■ Solving Algebraic Equations

`Solve[ls == rs, x]`	solve the equation $ls = rs$ for x
`Solve[{$g_1$, $g_2$, ...}, {$x_1$, $x_2$, ...}]`	solve a system of equations
`FindRoot[g, {x, $x_0$}]`	find a numerical root; the initial value is x_0

■ Calculus

`Limit[f, x -> $x_0$]`	the limit of f for $x \to x_0$
`D[f, x]`	the derivative of f with respect to x
`Integrate[f, x]`	the indefinite integral of f
`Integrate[f, {x, $x_{min}$, $x_{max}$}]`	the definite integral in the interval $[x_{min},\ x_{max}]$
`DSolve[`	solve the differential equation
`  x'[t] == x[t], x[t], t]`	$x'(t) = x(t)$ for $x(t)$
`NDSolve[`	find a numerical solution to the differential
`  {x'[t] == x[t], x[0] == 1},`	equation $x'(t) = x(t)$ with the initial
`  x[t], {t, $t_{min}$, $t_{max}$}]`	condition $x(0) = 1$ in the interval $[t_{min},\ t_{max}]$

▪ Plots

$\text{Plot}[f, \{x, x_{\min}, x_{\max}\}]$	plot a function of one variable
$\text{Plot3D}[f, \{x, x_{\min}, x_{\max}\}, \{y, y_{\min}, y_{\max}\}]$	plot a function of two variables
$\text{ContourPlot}[f, \{x, x_{\min}, x_{\max}\}, \{y, y_{\min}, y_{\max}\}]$	draw a contour plot of a function of two variables
$\text{ParametricPlot}[\{f_x, f_y\}, \{t, t_{\min}, t_{\max}\}]$	draw a parametric curve in the plane
$\text{ParametricPlot3D}[\{f_x, f_y, f_z\}, \{t, t_{\min}, t_{\max}\}]$	draw a parametric curve in space
$\text{ParametricPlot3D}[\{f_x, f_y, f_z\}, \{u, u_{\min}, u_{\max}\}, \{v, v_{\min}, v_{\max}\}]$	draw a parametric surface in space

▪ Lists and Matrices

$\text{Table}[f, \{i, i_{\min}, i_{\max}\}]$	create a list; the iterator i runs from i_{min} to i_{max} in increments of 1
$\text{Inverse}[m]$	the inverse of a matrix
$\text{Det}[m]$	the determinant of a matrix
$m \ . \ n$	matrix product

Part 1: The Basics

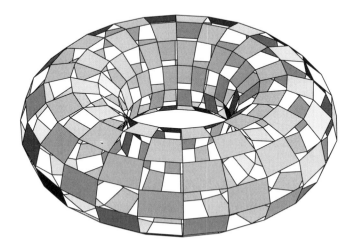

This part deals with the basics: the structure of the program, online documentation, input variations, as well as simple numerical calculations and symbolic mathematics.

■ 1.1 The Structure of the Program

Mathematica consists of two programs which can work independently of one another and even on separate computers. The two programs are called: *front end* and *kernel*.

■ 1.1.1 The Front End

Front end is the user-friendly interface with which *Mathematica* documents, called *notebooks*, can be created and edited. The many commands which are accessible via menus are documented in the *Help Browser* (menu **Help**) under **Other Information > Menu Commands**.

We launch the front end by double-clicking on the *Mathematica* icon (or using the `mathematica` command).

For actual calculations (after hitting the <Enter> key or SHIFT RET) the front end connects to the *kernel,* sends it the expressions to be evaluated, receives the results, and nicely displays them.

■ Cells and Styles

The front end arranges the notebooks into hierarchically grouped *cells*. Cells and their groupings are shown by the brackets on the right-hand side of the notebook. Cell groups can be opened and closed by double-clicking on the brackets, or with the **Cell > Cell Grouping** menu command. A new cell is created by clicking between two existing cells (or below the last cell) and typing the data to be entered.

Each cell has a style (menu **Format > Style**). The notebook uses predefined styles (**Format > Style Sheet**), which can be changed for every notebook using the style sheet or just for the current notebook (**Format > Edit Style Sheet...**). In the default notebook (**Default**) you can access, among other things, a hierarchy of title styles, text in two sizes, and styles for input and output cells.

Normally, if **Cell > Cell Grouping** is set to **Automatic Grouping**, the program organizes the cells automatically according to style by grouping cells between two titles, subtitles, etc., together.

■ **In Depth**

● **The Best Way to Organize Cells**

Text cells should contain one paragraph per cell.

Keeping each *Mathematica* expression in a single input cell gives a better overview of calculations. If necessary you can also combine several expressions divided by semicolons in one cell.

■ **Exercises**

● **Using the Front End**

Adjust the size of the notebook window to make it easier to read or to optimize the overview (menu **Format > Magnification**).

Open a new notebook (**File > New**).

Enter a title into a cell formatted for titles. You can either first choose the style (menu **Format > Style**) and then type, or first type (the cell will be formatted as *input*) and then select the cell bracket and change the style.

Below this, enter a section heading into a *section* cell. (Position the new cell by clicking below a cell or between two existing ones.)

Below this, enter text into a *text* cell.

Below this, enter a new section heading into a *section* cell.

Below this, enter a calculation (for example 1+1).

Evaluate the above cell using the $\boxed{\text{SHIFT}}$ $\boxed{\text{RET}}$ keys or by pressing <Enter>.

Note the automatic grouping of all cells.

Open and close some of the cell groups.

Save the notebook as a file (**File > Save As...**)

Select a different pre-defined style sheet (**Format > Style Sheet**) and note the change in appearance of the notebook.

■ 1.1.2 The Kernel

The *kernel* does the actual calculations. Normally you access it using the front end. It can also be launched by itself (by double-clicking on the *MathKernel* icon or using the `math` command).

When an input cell is evaluated using <Enter>, $\boxed{\text{SHIFT}}$ $\boxed{\text{RET}}$, or the **Evaluate Cells** command (menu **Kernel > Evaluation**) a kernel is launched and the automatic numbering of input and output cells begins at 1. During a kernel session, you will usually enter definitions (e.g., aConstant=3.1). These remain active until the end of the session if they are not explicitly cleared. Once you quit the kernel, all definitions are lost. You can reactivate them in the next session by evaluating the corresponding cells again.

Cells in a notebook can be evaluated in any order; it does not have to be from top to bottom. This may, however, give different results than if the notebook is evaluated sequentially (using for example the **Kernel > Evaluation > Evaluate Notebook**).

When you open a new notebook without quitting the front end, you continue to use the same kernel, which means that all definitions remain active. It is possible to configure additional kernels as necessary (**Kernel > Kernel Configuration Options > Add**) and to associate them to specific notebooks (**Kernel > Notebooks Kernel**). The kernels can also run on other computers.

When you quit the front end, the kernel processes stop automatically. Single evaluations can be aborted as necessary (**Kernel > Abort Evaluation**), or whole kernel processes can be terminated without quitting the front end (**Kernel > Quit Kernel**).

■ In Depth

• Communication between Front End and Kernel

The protocol used for the communication between the front end and the kernel is called *MathLink*. It can also be used to communicate between *Mathematica* and other application programs (see **Help > Add-Ons > MathLink Library**).

■ Exercises

• Starting and Aborting Evaluations

Start the following infinite loop:

```
While[True, 1]
```

Then abort it.

• Quitting the Kernel

Start the evaluation again.

Then quit the whole kernel process.

■ 1.2 Online Documentation

The **Help** menu contains several help commands. In addition to registration information and explanations of error beeps made by the computer (**Why the Beep?...**), you can open the *Help Browser* window to navigate through the online documentation. It is organized as follows:

Built-in Functions: organized by subject.

Add-ons: functions from packages that can be added to the program (see `Loading Packages`).

The Mathematica Book: the electronic version of the (1403 page) book by Stephen Wolfram. This is very useful, thanks to the hyperlinks.

Getting Started/Demos: various information and demonstrations. Have a look at it!

Other Information: front end menus, keyboard shortcuts.

Master Index: alphabetical index of all built-in functions.

You can access the different information using either the search function (enter text, select **Go To**), or by clicking on the hierarchically structured subject names. You also have the very useful option of selecting a function name in your notebook and accessing the documentation via the menu **Help > Find in Help...** (or > **Find Selected Function...**).

An incomplete installation of *Mathematica* can result in missing parts of the documentation (for example the book, which takes up a large portion of the hard drive).

■ Exercises

● Self Study

Open the *Help Browser*.

Study the organization of the **Built-in Functions**.

Note the underlined terms in the body text which indicate hyperlinks. After using a hyperlink, you can go back to your original location with the **Back** button or the menu **Find > Go Back**.

Have a look at the section "*Mathematica* as a Calculator" in the "Tour of *Mathematica*" (**Getting Started/Demos**).

Have a look at the subsection "Power Computing with *Mathematica*" of the section "Tour of *Mathematica*" in the *Mathematica Book*.

Study the documentation of the front end command **Find > Find...**.

Read the introduction to working with standard packages (**Add-ons > Working with Add-ons > Loading Packages**).

● **Packages**

The standard packages that come with the program contain many useful tools in addition to the functions built into the kernel. To use them you must first load the corresponding package.

Load the package `Miscellaneous`ChemicalElements``.

What is the atomic weight of plutonium?

■ 1.3 Formulas

■ 1.3.1 Formats

Input and output cells can basically be shown in three formats: `InputForm`, `Standard-Form`, and `TraditionalForm`.

In this course we use `InputForm` or `StandardForm`, depending on the circumstances.

■ **InputForm**

`InputForm` is useful for keyboard entries. (In early versions of *Mathematica* this was the only form of entry.) *Mathematica* functions are used by typing their names and placing their arguments inside square brackets. Further conventions are discussed below.

For example, the command for the integration of x (sin x) in `InputForm` looks like this:

> `Integrate[x Sin[x], x]`

The documentation of functions in the *Help Browser* is always shown in `InputForm`.

■ **StandardForm**

`StandardForm` is more similar to normal mathematical notation. It is unambiguous, unlike `TraditionalForm`. Integrals are written with the integral sign:

$$\int x \, \texttt{Sin[x]} \, dx$$

You can create input in `StandardForm` using either palettes or keyboard shortcuts–or by converting an `InputForm` cell to `StandardForm` (menu **Cell** > **Convert To** > **StandardForm**).

■ `TraditionalForm`

`TraditionalForm` follows the usual mathematical notation. The names of mathematical functions like "sin" are written in lower case, variables are in italics, and arguments are placed in round parentheses.

$$\int x \sin(x)\, d\, x$$

Unfortunately, this style contains many ambiguities which appear in mathematical texts, where they are resolved by the context or by implicit conventions. It is usually clear that the formula

$$a\,(b + c)$$

means the product of a multiplied by the sum of b and c. It is also normal to use

$$f(x)$$

for the application of the function f on the argument x. But how do we interpret the following formula:

$$f(b + c)$$

Is this the function f applied to the argument $b + c$ or the constant f multiplied by $b + c$? *Mathematica* cannot answer this question. For similar reasons, the special symbol d is used in integrals.

It is advantageous, therefore, to use only `InputForm` or `StandardForm` for input cells. Output cells can be generated in `TraditionalForm` as needed, either by converting the cell (**Cell** > **Convert To** > **TraditionalForm**), or by using the option **Cell** > **Default Output FormatType** > **TraditionalForm**.

■ Converting Cells

The following commands in the menu **Cell** are the most interesting for converting and displaying output cells :

Convert To: converts the selection to the format chosen.

Display As: displays the selection in the new format. Fractions, superscripts, etc. are not converted (unlike with **Convert To**).

Default Output FormatType: output cells are created in the format selected.

▪ Exercises

● Converting Formats

A derivative is written using the function name D in InputForm. The arguments are placed in square brackets and separated by commas. The first argument is the expression to be derived, the second is the variable to be used for the derivative:

```
D[Sin[2 x + a], x]
```

How is the derivative written in StandardForm or in TraditionalForm?

A second derivative looks like this in InputForm:

```
D[x Sin[x^3], {x, 2}]
```

Which are the other two forms of display?

■ 1.3.2 Entering Formulas and Special Characters

There are basically three methods for the easy entry of formulas and special characters. They can also be combined:
• using palettes,
• using control and escape key combinations,
• typing first in InputForm and subsequent conversion if necessary.

Mathematica contains a useful feature for working with formulas: a selection is enlarged hierarchically by repeatedly clicking on it.

▪ Palettes

The menu **File > Palettes** contains some useful pre-defined palettes.

AlgebraicManipulation: this is a compilation of several often-used functions for the algebraic manipulation of formulas, such as the expansion and factoring of polynomials and the simplification of expressions. Clicking the button on the palette automatically applies the function to the selection in the notebook, evaluating "on location".

BasicCalculations: contains the most important commands for simple calculations.

BasicInput: it makes sense to leave this palette on your screen. It contains the most-used symbols (Greek characters, etc.) and formulas (derivatives, integrals, etc.).

BasicTypesetting: an alternative or supplement to **Basic Input** containing many symbols, but no formulas.

CompleteCharacters: almost all special characters imaginable, organized by subject.

InternationalCharacters: this palette is useful if the needed international characters are not on your keyboard. It contains umlauts, etc.

NotebookLauncher: creates a new notebook with a chosen pre-defined style (analogous to the menu **Format > Style Sheet**).

Placeholders indicated by a ■ are filled out automatically with the current selection. The jump to the next placeholder can be shortened using the ⎄TAB key.

■ Control and Escape Key Shortcuts

Fractions, subscripts, etc. can also be created using the CTRL (<Control>) key in simultaneous combination with certain other keys. These shortcuts are shown in the menu **Edit > Expression Input**. The shortcut CTRL2 gives a square root whose radicand is entered automatically as you continue typing:

$$\sqrt{\square}$$

Many symbols can be written using escape sequences of the form ESC*key*ESC. You find the necessary keys in the **BasicTypesetting** palette by pointing at the desired symbol. To get Greek characters the analogous Latin key must be hit between the ESC keys. Typing ESCaESC therefore gives you an α.

Within nested formulas you can go back to the last level using CTRL␣ (<Control>- and spacebar). Therefore the key sequence CTRL/ a CTRL^ x CTRL␣ +b TAB c gives you the formula:

$$\frac{a^x + b}{c}$$

■ Using InputForm

As mentioned in the paragraph about formats, all input cells can also be written in the linear InputForm. If needed you can convert formulas into the two-dimensional StandardForm. In this case, roots and exponents look like this:

```
Sqrt[a] + b^3
```

After you select **Convert To > StandardForm** the cell becomes:

$$\sqrt{a} + b^3$$

Greek characters can also be entered using $\backslash$ [*name*]. If you replace *name* with `Alpha` you get an α.

`InputForm` and `StandardForm` formats can be combined with no problem:

$$\int \texttt{Sqrt[x] dx}$$

■ In Depth

● Creating Palettes

You can create your own palettes in three simple steps:
• select **Input > Create Palette**,
• fill out the palette and select it,
• select **File > Generate Palette from Selection**.
The ■ placeholder is created with ESC spl ESC. It will automatically be replaced by the current selection. A normal □ placeholder is written as ESC pl ESC.

In order for the palettes to appear in the menu **File > Palettes**, save the files in the subdirectory `Configuration\Front End\Palettes` of *Mathematica*'s installation directory or in the subdirectory `Front End/Palettes` of your personal *Mathematica* directory (for *Mathematica* 3.0 on a UNIX system, this would be: `~/.Mathematica/3.0`)

● Formulas Embedded in Text

As you see in this book, formulas can also be embedded in text cells. Here is an example: $\sqrt{x^2 + y^2}$. To achieve this, you can either copy an input or output cell which uses your favorite format (normally `TraditionalForm`) and paste it into the text cell. Or you open a placeholder box in the text cell with CTRL 9, use CTRL and ESC keystrokes to create the formula, and leave it with CTRL ⎵.

■ Exercises

● Self Study

Take a look at all the available palettes.

Study the keyboard shortcuts in the menu **Edit > Expression Input**.

● Writing Formulas

Create the following formula with three different methods: using palettes, using CTRL and ESC keys combinations (wherever possible), and by converting from `InputForm`.

$$\sqrt{\frac{\alpha}{3} + \sqrt{\frac{\beta}{2} + \sqrt{\gamma}}}$$

$$\int x^2 \, \text{Sin}[x] \, dx$$

$$\int_0^\pi \text{Sin}[x] \, \text{Cos}\left[x - \frac{\pi}{4}\right] dx$$

$$\frac{\partial^4 \frac{1}{\sqrt{x^2 + y^2}}}{\partial x^2 \, \partial y^2}$$

- **Hierarchy**

Click several times on one of the formulas created above and see how the selection is enlarged hierarchically.

- **Palettes**

Study the in-depth section "Creating Palettes". Then create your own palette. A simple example could be:

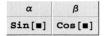

■ 1.4 Simple Calculations

Now we will begin with actual calculations. You can reproduce them on your computer using the SHIFT RET or <Enter> keys.

■ 1.4.1 Conventions

First we deal with the most important conventions in *Mathematica*. It is advisable to read this chapter quickly at first, and later, when you have made your own calculations, to study it more carefully.

■ Names

Mathematica is case sensitive.

In[1]:= **a - a**

Out[1]= 0

In[2]:= **a - A**

Out[2]= a - A

The names of built-in functions are (in InputForm) written with the first letter of each word capitalized. Each part of a compound word also begins with a capital.

In[3]:= **Expand[(a + b) ^2 / (c + d) ^2]**

Out[3]= $\dfrac{a^2}{(c+d)^2} + \dfrac{2\,a\,b}{(c+d)^2} + \dfrac{b^2}{(c+d)^2}$

In[4]:= **ExpandAll$\left[\dfrac{(a+b)^2}{(c+d)^2}\right]$**

Out[4]= $\dfrac{a^2}{c^2 + 2\,c\,d + d^2} + \dfrac{2\,a\,b}{c^2 + 2\,c\,d + d^2} + \dfrac{b^2}{c^2 + 2\,c\,d + d^2}$

To avoid conflicts between names of built-in *Mathematica* functions and other objects, you should begin your own names with a small letter.

In[5]:= **myFunction**

Out[5]= myFunction

In[6]:= **x**

Out[6]= x

Spaces (␣) can be used as long as they do not change the meaning of expressions.

In[7]:= **a -   a**

Out[7]= 0

But:

In[8]:= **aa / a**

Out[8]= $\dfrac{aa}{a}$

In[9]:= **a a / a**

Out[9]= a

(The space between the two a's indicates that the product a*a is meant–not the symbol named aa.)

▪ Parentheses, Brackets, and Braces

Arguments of *Mathematica* functions are placed in *square brackets* and separated by commas.

In[10]:= **Integrate[x^n, x]**

Out[10]= $\dfrac{x^{1+n}}{1+n}$

Parentheses are used for mathematical grouping.

In[11]:= **1 / (a + b (c + d))**

Out[11]= $\dfrac{1}{a + b \ (c + d)}$

Lists are placed in *curly braces*. They can be used, for instance, to define vectors. Lists are often also requested as arguments for built-in functions.

In[12]:= **{a, b, c}**

Out[12]= {a, b, c}

In[13]:= **Integrate[x^2, {x, 0, 1}]**

Out[13]= $\dfrac{1}{3}$

The *elements of lists* are numbered from left to right, starting with 1. *Double square brackets* (InputForm) or ⟦...⟧ brackets (StandardForm) are used to extract elements from lists.

In[14]:= **{a, b, c}[[1]]**

Out[14]= a

In[15]:= **{a, b, c}⟦2⟧**

Out[15]= b

Lists can also be *nested*:

In[16]:= `{{a, b, c}, {d, e, f}}`

Out[16]= `{{a, b, c}, {d, e, f}}`

To access a single element, we first indicate the position within the outer list, then the position in the corresponding sublist.

In[17]:= `{{a, b, c}, {d, e, f}}[[1, 2]]`

Out[17]= `b`

In[18]:= `{{a, b, c}, {d, e, f}}[[2, 3]]`

Out[18]= `f`

■ References to Results

Mathematica's input and output cells are automatically numbered in the order of their evaluation (In[...], Out[...]). The expression %*n* is a short form for the output cell with the number *n* (i.e.: Out[*n*]). % indicates the last output cell, %% indicates the one before the last, etc.

In[19]:= `2 %`

Out[19]= `2 f`

In[20]:= `% * %17`

Out[20]= `2 b f`

■ The Order of Evaluation

The order of evaluation does not need to be from top to bottom; cells may also be evaluated several times. In this case, however, once the notebook has been saved and evaluated in a new kernel, the results can be different if the order of definitions (see Section 1.4.4) has changed or if references to output cells are no longer correct.

■ Suppressing or Shortening the Output

If you add `;` to the end of an expression, *Mathematica* suppresses the display of the output. It gets evaluated nonetheless:

In[21]:= **a ^ 2;**

In[22]:= **%**

Out[22]= a^2

This is useful for calculations with huge results where the formatting of an output of several pages takes a lot of time. Shortened results can be created with Short or Shallow.

In[23]:= **Expand[(a + b + c) ^ 100];**

In[24]:= **Short[%]**

Out[24]//Short=
$$a^{100} + 100\, a^{99}\, b + \ll 5147 \gg + 100\, b\, c^{99} + c^{100}$$

- ### In Depth

- ### Notations

In addition to *standard notation*

In[25]:= **Expand[(a + b) ^2]**

Out[25]= $a^2 + 2\,a\,b + b^2$

functions with one argument can be written in a *prefix notation* using @

In[26]:= **Expand @ ((a + b) ^2)**

Out[26]= $a^2 + 2\,a\,b + b^2$

or in a *postfix notation* using //

In[27]:= **(a + b) ^2 // Expand**

Out[27]= $a^2 + 2\,a\,b + b^2$

For functions with two arguments you can also use *infix notation*:

In[28]:= **{a, b} ~ Join~ {c, d}**

Out[28]= $\{a, b, c, d\}$

■ 1.4.2 Numerical Calculations

The operators for addition (+), subtraction (−), multiplication (∗), division (/), and powers (^) are the usual ones. The multiplication asterisk can also be replaced by a space.

In[29]:= **2 3 / 5**

Out[29]= $\dfrac{6}{5}$

Mathematica works with exact integers or rational numbers, as long as there is no decimal point.

In[30]:= **2 ^ 100**

Out[30]= 1267650600228229401496703205376

In[31]:= **2.0 ^ 100**

Out[31]= 1.26765×10^{30}

In[32]:= **$\sqrt{2}$**

Out[32]= $\sqrt{2}$

The conversion to approximate numbers is done by the function N.

In[33]:= **N$\left[\sqrt{2}\,\right]$**

Out[33]= 1.41421

An optional second argument demands greater precision.

In[34]:= **N$\left[\sqrt{2}\,,\ 50\right]$**

Out[34]= 1.4142135623730950488016887242096980785696718753769

Mathematica also recognizes various constants, e.g.:

E or *e*: $e \approx 2.71828$
Pi or π: $\pi \approx 3.14159$
I or *i*: $i = \sqrt{-1}$
Degree: $\pi/180$, the number of radians in one degree

As long as a numerical approximation is not requested, these constants are used as purely symbolic expressions. Certain properties are (exactly) known.

In[35]:= $\dfrac{\pi}{4}$

Out[35]= $\dfrac{\pi}{4}$

In[36]:= **Sin** $\left[\dfrac{\pi}{4} \right]$

Out[36]= $\dfrac{1}{\sqrt{2}}$

In[37]:= **N[Pi / 4, 20]**

Out[37]= 0.78539816339744830962

In[38]:= **Sin[45 Degree]**

Out[38]= $\dfrac{1}{\sqrt{2}}$

The many built-in mathematical functions and constants can best be found in the *Help Browser* (under **Built-in Functions > Elementary Functions**) or in the **BasicCalculations** palette. Their numerical evaluation is simple:

In[39]:= **ArcCos[0]**

Out[39]= $\dfrac{\pi}{2}$

In[40]:= **ArcCos[7 / 10]**

Out[40]= $\text{ArcCos}\left[\dfrac{7}{10} \right]$

In[41]:= **N[%]**

Out[41]= 0.795399

In[42]:= **ArcCos[.7]**

Out[42]= 0.795399

■ Exercises

• The Exponential Constant

Have a look at the first 1000 places of e.

• Approximations

Determine the absolute and the relative error of the approximation of π by the square root of 10.

■ 1.4.3 Algebraic Manipulation

Mathematica can also handle symbols.

In[43]:= **(a + b) ^ 10**

Out[43]= $(a + b)^{10}$

Only the simplest calculations are carried out automatically. All others must be requested specifically, since the program cannot know what we want to do with a formula.

To expand out the above polynomial we can use several methods. We can find the applicable function Expand in the *Help Browser* (**Built-in Functions** > **Algebraic Manipulation** > **Basic Algebra**) and type it into the notebook. The % sign is used to reference the last output.

In[44]:= **Expand[%]**

Out[44]= $a^{10} + 10\,a^9\,b + 45\,a^8\,b^2 + 120\,a^7\,b^3 + 210\,a^6\,b^4 +$
$252\,a^5\,b^5 + 210\,a^4\,b^6 + 120\,a^3\,b^7 + 45\,a^2\,b^8 + 10\,a\,b^9 + b^{10}$

Or we make a copy of $(a+b)^{\wedge}10$, select it, and use the palette **BasicCalculations** > **Algebra** > **Polynomial Manipulations** to click **Expand[■]** into the notebook. The placeholder ■ will automatically be replaced by the selection.

In[45]:= **Expand[(a + b) ^ 10]**

Out[45]= $a^{10} + 10\,a^9\,b + 45\,a^8\,b^2 + 120\,a^7\,b^3 + 210\,a^6\,b^4 +$
$252\,a^5\,b^5 + 210\,a^4\,b^6 + 120\,a^3\,b^7 + 45\,a^2\,b^8 + 10\,a\,b^9 + b^{10}$

As an alternative we can also select the formula and apply the function Expand[■] from the palette **AlgebraicManipulation**. The cell is evaluated "on location" and

(a + b) ^ 10

changes into:

$a^{10} + 10\,a^9\,b + 45\,a^8\,b^2 + 120\,a^7\,b^3 + 210\,a^6\,b^4 +$
$252\,a^5\,b^5 + 210\,a^4\,b^6 + 120\,a^3\,b^7 + 45\,a^2\,b^8 + 10\,a\,b^9 + b^{10}$

Try this yourself.

One of the most favorite functions is Simplify. When we apply it to the above expanded polynomial, *Mathematica* returns it in its factored form, which clearly is much simpler.

In[47]:= **Simplify[a^{10} + 10 a^9 b + 45 a^8 b^2 + 120 a^7 b^3 + 210 a^6 b^4 +**
      252 a^5 b^5 + 210 a^4 b^6 + 120 a^3 b^7 + 45 a^2 b^8 + 10 a b^9 + b^{10}]

Out[47]= $(a + b)^{10}$

In this case, Factor produces the same result.

In[48]:= **Factor[a^{10} + 10 a^9 b + 45 a^8 b^2 + 120 a^7 b^3 + 210 a^6 b^4 +**
      252 a^5 b^5 + 210 a^4 b^6 + 120 a^3 b^7 + 45 a^2 b^8 + 10 a b^9 + b^{10}]

Out[48]= $(a + b)^{10}$

The function FullSimplify often takes longer than Simplify, but it recognizes additional (and sometimes quite exotic) rules:

In[49]:= **Simplify$\left[$ArcCos$\left[\sqrt{1-x}\,\right]\right]$**

Out[49]= ArcCos$[\sqrt{1-x}\,]$

In[50]:= **FullSimplify$\left[$ArcCos$\left[\sqrt{1-x}\,\right]\right]$**

Out[50]= ArcSin$[\sqrt{x}\,]$

The simplification of formulas is a difficult problem which (in general) must be approached heuristically. The difficulties already start with the concept. Which of the following formulas is simpler?

In[51]:= **FullSimplify$\left[\dfrac{1 - x^{11}}{1 - x}\right]$**

Out[51]= $\dfrac{1 - x^{11}}{1 - x}$

In[52]:= **FullSimplify[1 + x + x^2 + x^3 + x^4 + x^5 + x^6 + x^7 + x^8 + x^9 + x^{10}]**

Out[52]= $1 + x (1 + x (1 + x + x^2) (1 + x^3 + x^6))$

It is debatable. We therefore will not hold it against *Mathematica* for not using the same form in both cases, although they seem to be identical.

In[53]:= **Simplify[% - %%]**

Out[53]= 0

(Are they really the same?)

■ Exercises

● A Simplification

Use the appropriate function from the **Basic Calculations** palette (or type the name of the function) to simplify the following expression:

$$a^5 + 5\,a^4\,\text{Cos}[x]^2 + 10\,a^3\,\text{Cos}[x]^4 + 10\,a^2\,\text{Cos}[x]^6 +$$
$$5\,a\,\text{Cos}[x]^8 + \text{Cos}[x]^{10} + 5\,a^4\,\text{Sin}[x]^2 + 20\,a^3\,\text{Cos}[x]^2\,\text{Sin}[x]^2 +$$
$$30\,a^2\,\text{Cos}[x]^4\,\text{Sin}[x]^2 + 20\,a\,\text{Cos}[x]^6\,\text{Sin}[x]^2 + 5\,\text{Cos}[x]^8\,\text{Sin}[x]^2 +$$
$$10\,a^3\,\text{Sin}[x]^4 + 30\,a^2\,\text{Cos}[x]^2\,\text{Sin}[x]^4 + 30\,a\,\text{Cos}[x]^4\,\text{Sin}[x]^4 +$$
$$10\,\text{Cos}[x]^6\,\text{Sin}[x]^4 + 10\,a^2\,\text{Sin}[x]^6 + 20\,a\,\text{Cos}[x]^2\,\text{Sin}[x]^6 +$$
$$10\,\text{Cos}[x]^4\,\text{Sin}[x]^6 + 5\,a\,\text{Sin}[x]^8 + 5\,\text{Cos}[x]^2\,\text{Sin}[x]^8 + \text{Sin}[x]^{10}$$

● Calculating "on Location"

Use the **AlgebraicManipulation** palette to:
• expand out $(a + b)^{10}$,
• factor the result,
• simplify `Sin[2α+β] Cos[2α+β]`,
• simplify `Log[z + √z + 1 √z - 1]`
  (compare the results of `Simplify` and `FullSimplify`).

● Self Study

Take a look at Section 1.4.5 in the *Mathematica Book*.

● Goniometric Relationships

Convert the formula

`Sin[3 x] Cos[5 x]`

into a form in which no multiples of x appear in the trigonometric functions.

■ 1.4.4 Transformation Rules and Definitions

This section will be difficult on first reading. Read it through first and return to it later on, whenever the use of transformation rules or definitions is unclear to you.

■ Transformation Rules

Replacing values for symbols is done by the operator `/.` in which a *transformation rule* has to be given on the right-hand side. The latter is written in `InputForm` as *variable* `->` *value* or in `StandardForm` as *variable* $\rightarrow$ *value*.

In[54]:= **Sqrt[a + b^2] /. a -> 2**

Out[54]= $\sqrt{2 + b^2}$

In[55]:= **$\sqrt{a + b^2}$ /. b → 3**

Out[55]= $\sqrt{9 + a}$

Several simple rules can be combined in a list.

In[56]:= **$\sqrt{a + b^2}$ /. {a → 3, b → 7}**

Out[56]= $2\sqrt{13}$

We also call such a list *transformation rule* because it acts like a single transformation rule–unlike a nested list. Nested lists allow us to substitute different values at the same time.

In[57]:= **$\sqrt{a + b^2}$ /. {{a → c, b → 0}, {a → a^2}}**

Out[57]= $\left\{ \sqrt{c}, \sqrt{a^2 + b^2} \right\}$

■ Simple Definitions

An *immediate definition* is indicated by an equals sign (=).

In[58]:= **a1 = 1**

Out[58]= 1

The right-hand side of the immediate definition is evaluated when the definition is evaluated: You can see the result in the output cell. During the *Mathematica* session, the definition is applied whenever the left-hand side of the definition matches a subexpression.

In[59]:= **a1 + a2**

Out[59]= 1 + a2

In[60]:= **a2 = a1**

Out[60]= 1

In[61]:= **a2**

Out[61]= 1

Delayed definitions (:=) are also used wherever their left-hand side appears. But in this case the evaluation of the right-hand side is delayed until the definition is used. Therefore we do not get an output cell.

In[62]:= **a3 := a1**

In[63]:= **a3**

Out[63]= 1

If we change the value of a1 and evaluate a3 again, we get a different result.

In[64]:= **a1 = 3**

Out[64]= 3

In[65]:= **a3**

Out[65]= 3

The value of a2, which was set with an immediate definition, has not changed.

In[66]:= **a2**

Out[66]= 1

We can look at the definitions associated to *name* by evaluating ?*name*.

In[67]:= **?a3**

    Global`a3

    a3 := a1

This shows us that a3 is in the context **Global`** (see Section 4.4.5) and that the definition a3:=a1 has been set for it.

■ Clearing Definitions

Immediate and delayed definitions are cleared with Clear or =..

In[68]:= **Clear[a2, a3]**

In[69]:= **{a1, a2, a3}**

Out[69]= {3, a2, a3}

In[70]:= **a1 =.**

In[71]:= **{a1, a2, a3}**

Out[71]= {a1, a2, a3}

▪ Simple Patterns

The left-hand side of transformation rules and definitions are actually *patterns.* Up to now, they have been very simple, because they contained only single symbol names. But wherever there is a *blank* (_ , [SHIFT]- on the keyboard) in a pattern any expression can appear in its place. Therefore the blank symbol "_" stands for "anything".

In[72]:= **1 + a ^ 2 /. _ ^ 2 -> somethingSquared**

Out[72]= 1 + somethingSquared

In definitions we usually need this "anything" on the right-hand side. It can therefore be associated with a name. So x_ stands for any given number, symbol, or more general expression which will be referenced by the name x on the right-hand side. We can use this to define *functions*:

In[73]:= **function1[x_] = Sin[1 / x]**

Out[73]= $Sin\left[\frac{1}{x}\right]$

In[74]:= **function1[3]**

Out[74]= $Sin\left[\frac{1}{3}\right]$

In[75]:= **function1[f[Tan[a]]]**

Out[75]= $Sin\left[\frac{1}{f[Tan[a]]}\right]$

In this example a delayed definition would have given the same result. But if something needs to be evaluated on the right, then it makes a difference what type of definition it is. Look at the two following definitions:

In[76]:= **myExpand1[x_] = Expand[(1 + x) ^ 2]**

Out[76]= $1 + 2x + x^2$

In[77]:= **myExpand2[x_] := Expand[(1 + x) ^ 2]**

When applied to a single symbol or a number they give the same result.

In[78]:= **myExpand1[a]**

Out[78]= $1 + 2\,a + a^2$

In[79]:= **myExpand2[a]**

Out[79]= $1 + 2\,a + a^2$

But if we evaluate them for a sum, this sum is simply substituted for x in the first version.

In[80]:= **myExpand1[a + b]**

Out[80]= $1 + 2\,(a + b) + (a + b)^2$

On the other hand, with a delayed definition the sum is substituted and the Expand of the resulting expression is then calculated.

In[81]:= **myExpand2[a + b]**

Out[81]= $1 + 2\,a + a^2 + 2\,b + 2\,a\,b + b^2$

■ Rules of Thumb for Definitions

We can keep to the rule of thumb that immediate definitions serve as shortcuts for *fixed values* of symbols or patterns. But if something needs to be *calculated* when the definition is used, then a delayed definition is appropriate.

Because definitions are valid throughout a *Mathematica* session if they are not cleared by hand, they can lead to confusion if you are working on a larger project and forget them. Transformation rules are therefore more suitable for substituting values.

■ In Depth

● Clearing All Definitions

This clears all definitions without launching a new kernel session (see Section 4.4.5):

In[82]:= Clear[Global`*]

● Compound Expressions

If needed, we can combine several expressions on one line or in one cell by separating them with semicolons. This is called a *compound expression.*

In[83]:= **const1 = .2; const2 = .3; {const1, const2}**

Out[83]= {0.2, 0.3}

▪ Exercises

● Substituting Values

In the following expression, first have a=2, then b=3 (with any value for a), and then at the same time have a=2 and b=3.

$$\frac{a^2 - b}{b^3 + a^2 + x}$$

● A Function Definition

Define a function with two arguments n and x, which calculates $\sin(n\,x)$.

● Stirling's Formula

For large n Stirling's formula is valid: $\log n! \approx \left(n + \frac{1}{2}\right) \log n - n + \log \sqrt{2\pi}$.

Calculate the absolute and the relative errors for $n = 2$, 10, 100. First use transformation rules and then definitions.

▪ 1.4.5 Equations

▪ Single Equations

Equations (and differential equations) are indicated in *Mathematica* with a double equals sign, since the simple equals sign is already taken by definitions.

In[84]:= **a x + b == 1**

Out[84]= b + a x == 1

In[85]:= **myEquation = a x + b == 1**

Out[85]= b + a x == 1

The *Mathematica* function for solving one or more equations is called Solve. It needs to know the equation and the variable.

In[86]:= **Solve[a x + b == 1, x]**

Out[86]= $\left\{\left\{x \rightarrow -\frac{-1 + b}{a}\right\}\right\}$

In[87]:= **Solve[myEquation, x]**

Out[87]= $\left\{\left\{x \to -\frac{-1+b}{a}\right\}\right\}$

Let us give the solution a name:

In[88]:= **result = %**

Out[88]= $\left\{\left\{x \to -\frac{-1+b}{a}\right\}\right\}$

The result of Solve is written as a list of transformation rules, which may be irritating at first. In addition, it is a nested list, because we can also solve sets of equations with several solutions and several unknowns. We get the first (and in this case only) solution by accessing the first (and only) element of the list:

In[89]:= **firstSolution = result[[1]]**

Out[89]= $\left\{x \to -\frac{-1+b}{a}\right\}$

This is a transformation rule which we can apply to expressions. We substitute the solution for x into the equation with:

In[90]:= **myEquation /. firstSolution**

Out[90]= True

The answer to the frequently asked question, how to set x to this value definitively, is:

In[91]:= **x = x /. firstSolution**

Out[91]= $-\frac{-1+b}{a}$

In[92]:= **x + 1**

Out[92]= $1 - \frac{-1+b}{a}$

or in one step:

In[93]:= **x = x /. Solve[a x + b == 1, x][[1]]**

General::ivar : $-\dfrac{-1+b}{a}$ is not a valid variable.

ReplaceAll::reps :
  {True} is neither a list of replacement rules nor a valid
    dispatch table, and so cannot be used for replacing.

Out[93]= $-\dfrac{-1+b}{a}$ /. True

This already shows us the danger in this kind of definition: x already has a value through the definition x = x /. firstSolution. This is immediately substituted into the equation which evaluates to True. The name x can therefore no longer be used as a variable. It is better if we delete the definition for now

In[94]:= **x =.**

and avoid the definition for x:

In[95]:= **x /. Solve[a x + b == 1, x][[1]]**

Out[95]= $-\dfrac{-1+b}{a}$

Nonlinear equations are more interesting:

In[96]:= **threeSolutions = Solve[x^3 + x^2 - x + 1 == 0, x]**

Out[96]= $\left\{ \left\{ x \to -\dfrac{1}{3} - \dfrac{4}{3\,(19-3\,\sqrt{33})^{1/3}} - \dfrac{1}{3}\,(19-3\,\sqrt{33})^{1/3} \right\}, \right.$

$\left\{ x \to -\dfrac{1}{3} + \dfrac{2\,(1+I\,\sqrt{3})}{3\,(19-3\,\sqrt{33})^{1/3}} + \dfrac{1}{6}\,(1-I\,\sqrt{3})\,(19-3\,\sqrt{33})^{1/3} \right\},$

$\left. \left\{ x \to -\dfrac{1}{3} + \dfrac{2\,(1-I\,\sqrt{3})}{3\,(19-3\,\sqrt{33})^{1/3}} + \dfrac{1}{6}\,(1+I\,\sqrt{3})\,(19-3\,\sqrt{33})^{1/3} \right\} \right\}$

This gives us three solutions. The list of transformation rules can be applied as a whole to an expression. The result is the list of the three substitutions.

In[97]:= **x /. threeSolutions**

Out[97]= $\left\{ -\dfrac{1}{3} - \dfrac{4}{3\,(19 - 3\,\sqrt{33}\,)^{1/3}} - \dfrac{1}{3}\,(19 - 3\,\sqrt{33}\,)^{1/3} , \right.$

$\qquad -\dfrac{1}{3} + \dfrac{2\,(1 + I\,\sqrt{3}\,)}{3\,(19 - 3\,\sqrt{33}\,)^{1/3}} + \dfrac{1}{6}\,(1 - I\,\sqrt{3}\,)\,(19 - 3\,\sqrt{33}\,)^{1/3} ,$

$\qquad \left. -\dfrac{1}{3} + \dfrac{2\,(1 - I\,\sqrt{3}\,)}{3\,(19 - 3\,\sqrt{33}\,)^{1/3}} + \dfrac{1}{6}\,(1 + I\,\sqrt{3}\,)\,(19 - 3\,\sqrt{33}\,)^{1/3} \right\}$

Let us verify the solution.

In[98]:= **Simplify[x^3 + x^2 - x + 1 == 0 /. threeSolutions]**

> $MaxExtraPrecision::meprec :
> In increasing internal precision while attempting to evaluate
> $\dfrac{4}{3} + \dfrac{4}{3\,(\ll 1 \gg)^{1/3}} + \dfrac{1}{3}\,(\ll 1 \gg)^{1/3} + (\ll 1 \gg)^2 + (\ll 1 \gg)^3$, the
> limit $MaxExtraPrecision = 50.` was reached. Increasing the
> value of $MaxExtraPrecision may help resolve the uncertainty.

> $MaxExtraPrecision::meprec :
> In increasing internal precision while attempting to evaluate
> $\dfrac{4}{3} - \dfrac{2\,(1 + I\,\sqrt{3}\,)}{3\,(\ll 1 \gg)^{1/3}} - \ll 1 \gg + (\ll 1 \gg)^2 + (\ll 1 \gg)^3$, the limit
> $MaxExtraPrecision = 50.` was reached. Increasing the value
> of $MaxExtraPrecision may help resolve the uncertainty.

> $MaxExtraPrecision::meprec :
> In increasing internal precision while attempting to evaluate
> $\dfrac{4}{3} - \dfrac{2\,(1 - I\,\sqrt{3}\,)}{3\,(\ll 1 \gg)^{1/3}} - \ll 1 \gg + (\ll 1 \gg)^2 + (\ll 1 \gg)^3$, the limit
> $MaxExtraPrecision = 50.` was reached. Increasing the value
> of $MaxExtraPrecision may help resolve the uncertainty.

> General::stop :
> Further output of $MaxExtraPrecision::meprec will be
> suppressed during this calculation.

Out[98]= {True, True, True}

The result is correct; the messages (produced by Version 3.0.1) should not appear. Like every nontrivial program, *Mathematica* is not perfect. Just to double check we will try an alternate version, where we only calculate the left-hand side of the equation.

In[99]:= **Simplify[x^3 + x^2 - x + 1 /. threeSolutions]**

Out[99]= {0, 0, 0}

■ Sets of Equations

In order to solve a set of simultaneous equations, we group the equations and the unknowns as lists.

In[100]:= **Solve[{2 x^2 + y == 1, x - y == 2}, {x, y}]**

Out[100]= $\left\{\left\{y \to -\frac{7}{2},\ x \to -\frac{3}{2}\right\},\ \{y \to -1,\ x \to 1\}\right\}$

We thereby finally recognize the meaning of the display of solutions as nested lists. We have a list with two solutions and each solution is a list of rules for the two unknowns.

In[101]:= **Simplify[{2 x^2 + y == 1, x - y == 2} /. %]**

Out[101]= {{True, True}, {True, True}}

The Eliminate function is sometimes also useful. It eliminates variables from a set of equations.

In[102]:= **Eliminate[{x - y == d, x + y == s}, x]**

Out[102]= d == s - 2 y

■ Numerical Solutions of Polynomial Equations

The solutions of polynomial equations of degree > 4 can generally not be written as rational expressions with radicals.

In[103]:= **Solve[x^5 - x^2 + 1 == 0, x]**

Out[103]= {{x → Root[1 - #1^2 + #1^5 &, 1]},
{x → Root[1 - #1^2 + #1^5 &, 2]}, {x → Root[1 - #1^2 + #1^5 &, 3]},
{x → Root[1 - #1^2 + #1^5 &, 4]}, {x → Root[1 - #1^2 + #1^5 &, 5]}}

We do not want to get further into Root objects (with which you can also calculate), rather we want to create a numerical approximation of the solutions.

In[104]:= **N[%]**

Out[104]= {{x → -0.808731}, {x → -0.464912 - 1.07147 I},
{x → -0.464912 + 1.07147 I},
{x → 0.869278 - 0.388269 I}, {x → 0.869278 + 0.388269 I}}

Aside from numeric subtleties, the function NSolve gives us the same result as N[Solve[...]]

In[105]:= **NSolve[x⁵ - x² + 1 == 0, x]**

Out[105]= {{x → -0.808731}, {x → -0.464912 - 1.07147 I},
           {x → -0.464912 + 1.07147 I},
           {x → 0.869278 - 0.388269 I}, {x → 0.869278 + 0.388269 I}}

■ Numerical Solutions of Transcendental Equations

Unfortunately, there are also transcendental equations which can possibly have several or an infinite number of solutions.

In[106]:= **Solve[Log[x] == Cot[x], x]**

          Solve::tdep :
            The equations appear to involve transcendental functions
              of the variables in an essentially non-algebraic way.

Out[106]= Solve[Log[x] == Cot[x], x]

With a look ahead at Part 2, let us create at a plot of both sides of the equation.

In[107]:= **Plot[{Log[x], Cot[x]}, {x, 0, 4 π}]**

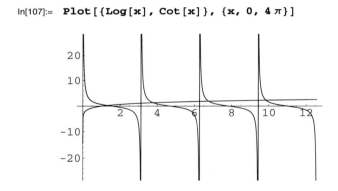

Out[107]= - Graphics -

This shows us that there are (infinitely) many solutions. We can only use a numerical algorithm to search an approximation of one solution (see also Section 4.4.2). The function that does this for us, FindRoot, demands that the equation (or the expression whose roots are being sought) be the first argument and that a list with the variable and the initial value be the second.

In[108]:= **FindRoot[Log[x] == Cot[x], {x, 1}]**

Out[108]= {x → 1.30828}

A different initial value may deliver a different solution.

In[109]:= **FindRoot[Log[x] == Cot[x], {x, 7}]**

Out[109]= $\{x \rightarrow 6.76512\}$

Note that the solution found by the numerical algorithm does not need to be the closest to the starting value. Here we use expressions instead of equations.

In[110]:= **FindRoot[x^4 - 2 x^2 + 1 / 2, {x, .1}]**

Out[110]= $\{x \rightarrow 1.30656\}$

In[111]:= **FindRoot[x^4 - 2 x^2 + 1 / 2, {x, .2}]**

Out[111]= $\{x \rightarrow 0.541196\}$

You can find further ways of calling up FindRoot in the *Help Browser*.

▪ In Depth

● Special Cases

Let us look at the solution of $a\,x = 1$.

In[112]:= **Solve[a x == b, x]**

Out[112]= $\left\{\left\{x \rightarrow \dfrac{b}{a}\right\}\right\}$

Obviously, if $b \neq 0$, the solution is not valid for $a = 0$.

In[113]:= **%[[1]] /. a -> 0**

Power::infy : Infinite expression $\dfrac{1}{0}$ encountered.

Out[113]= $\{x \rightarrow \text{ComplexInfinity}\}$

We see here, that the function Solve does not take special cases into consideration. Technically speaking, it only delivers a *generic solution*.

The Reduce function helps us further. It creates a logical expression containing all special cases.

In[114]:= **Reduce[a x == b, x]**

Out[114]= $b == 0 \,\&\&\, a == 0 \,||\, a \neq 0 \,\&\&\, x == \dfrac{b}{a}$

The logical *or* is written | |, the logical *and* as &&, and *unequal* as ≠ or ! = .

• Inequalities

The `InequalitySolve` function is defined in the `Algebra`InequalitySolve`` package. It helps us simplify inequalities. To use it we must first load the package.

  In[115]:= `<< Algebra`InequalitySolve``

Now we can simplify the following inequality, for example:

  In[116]:= `InequalitySolve[x^2 - 3 > 0, x]`

Out[116]= $x < -\sqrt{3} \;||\; x > \sqrt{3}$

Hence, x must satisfy either $x < -\sqrt{3}$ or $x > \sqrt{3}$.

■ Exercises

• Quadratic Equations

Solve the quadratic equation $a\,x^2 + b\,x + c = 0$ for x.

Verify the result by substituting it into the equation.

Define a variable with the name `solution1`, whose value is the first solution of the equation.

Find a form of the solution that also takes special cases like $a = 0$ into account.

• Equations of Higher Degree

Study the symbolic solutions to the equation:

$$4\,x^4 + 3\,x^3 + 2\,x^2 + x + 1 == 0$$

Create numerical approximations of the solutions using different methods.

• Transcendental Equations

Find the first two positive points of intersections of e^{-x} and $\sin(x)$.

• Sets of Simultaneous Equations

Solve the following set and verify the solutions by substituting them into the equations.

$$\{x^2 + y == 1,\; 3\,y - x == a\}$$

• Elimination of Variables

Eliminate x and y from the following set of simultaneous equations:

$$\{x^2 + y + z == 1,\; 3\,y - x == a,\; x + 2\,z == 0\}$$

• Inequalities

Determine where the inequalities $|x^2 - 3| - 2 > 0$ and $x^2 - x^3 > 0$ hold (separately and simultaneously).

■ 1.4.6 Calculus

■ Limits

Limits are determined with the function `Limit` as follows:

In[117]:= **Limit[(x - 1)^2 / (x^2 - 1), x -> 1]**

Out[117]= 0

In[118]:= **Limit$\left[\dfrac{(x - 1)^2}{x^2 - 1}, x \to -1\right]$**

Out[118]= $-\infty$

We see that the symbol `Infinity` or ∞ is predefined. We can use it for limits or integrals, among other things.

In[119]:= **Limit[Log[x] / x, x -> Infinity]**

Out[119]= 0

For the above expression $\frac{(x-1)^2}{x^2-1}$ the limits are different if x approaches -1 from smaller values or from larger values.

In[120]:= **Plot$\left[\dfrac{(x - 1)^2}{x^2 - 1}, \{x, -2, 2\}\right]$**

Out[120]= - Graphics -

The option `Direction` can be used to differentiate between the limit from the left (`Direction→1`) and the limit from the right (`Direction→-1`).

In[121]:= **Limit** $\left[\dfrac{(x-1)^2}{x^2-1}, \ x \to -1, \ \textbf{Direction} \to 1\right]$

Out[121]= ∞

In[122]:= **Limit** $\left[\dfrac{(x-1)^2}{x^2-1}, \ x \to -1, \ \textbf{Direction} \to -1\right]$

Out[122]= $-\infty$

Many other *Mathematica* functions can be manipulated in an analog way using *options*. These are always written as transformation rules. In Part 2 we will see many further examples with graphic functions.

■ Derivatives

We have already seen the function D for the calculation of derivatives. Because it is used often, its name is (like N) one of the few exceptions in naming where only a letter is used in place of a whole word.

In InputForm the expression to be derived comes first, then the variable or a list containing the variable and the multiplicity of the derivative.

In[123]:= **D[x^2, x]**

Out[123]= $2\,x$

In[124]:= **D[Sin[x], {x, 2}]**

Out[124]= $-\text{Sin}[x]$

In StandardForm (see palette **BasicCalculations > Calculus > Common Operations**) the input cells are written a little differently:

In[125]:= $\partial_x\, x^2$

Out[125]= $2\,x$

In[126]:= $\partial_{\{x,2\}}\, \textbf{Sin[x]}$

Out[126]= $-\text{Sin}[x]$

Let us calculate the derivative of an unknown function.

In[127]:= **D[f[x], x]**

Out[127]= f' [x]

Apostrophes may also be used to enter derivatives of functions of one variable. This is particularly useful for differential equations. *Mathematica* treats both variations identically.

In[128]:= **f ' [x] - %**

Out[128]= 0

■ Integrals

We use the function name Integrate (in InputForm) or the palette **BasicCalcula-tions > Calculus > Common Operations** for calculating integrals.

In[129]:= **Integrate[x Sin[x], x]**

Out[129]= -x Cos [x] + Sin [x]

In[130]:= $\int$ **x Cos [x] d x**

Out[130]= Cos [x] + x Sin [x]

Mathematica sets the constant of integration in indefinite integrals to zero.

As you have probably been starting to guess, the variable and the end points of a definite integral must be given as a list.

In[131]:= **Integrate[x Log[x], {x, a, b}]**

Out[131]= $-\frac{1}{4}$ a^2 (-1 + 2 Log [a]) + $\frac{1}{4}$ b^2 (-1 + 2 Log [b])

The entry is even easier using the palette. Click on the template

$$\int_{\square}^{\square} \square \, d\square$$

and jump from placeholder to placeholder using the tab key. This gives us for example:

In[132]:= $\int_{0}^{2\pi}$ **(a - a Cos [t])^2 d t**

Out[132]= 3 a^2 π

Integrals of expressions with elementary functions are–unlike their derivatives–often no longer elementary. Either the results are special functions that are basically defined as being the integral of another function

In[133]:= $\int \text{Exp}[x^2]\, dx$

Out[133]= $\frac{1}{2} \sqrt{\pi}\, \text{Erfi}[x]$

or the integral is returned unevaluated:

In[134]:= $\int_1^2 \text{Exp}[x^2]\, \text{Log}[x^2]\, \text{Sin}[x^2]\, dx$

Out[134]= $\int_1^2 E^{x^2}\, \text{Log}[x^2]\, \text{Sin}[x^2]\, dx$

Mathematica does not calculate integrals in the same way you learned in school. The *Risch* algorithm implemented in the program can calculate an entire class of integrals and can also decide whether the result exists as a function in this class. In addition, *Mathematica* recognizes many definite integrals which can be written as hypergeometric or other special functions.

The function NIntegrate returns numerical approximations of definite integrals.

In[135]:= **NIntegrate[Exp[x²] Log[x²] Sin[x²], {x, 1, 2}]**

Out[135]= -2.22919

■ Differential Equations

Analogous to algebraic equations, differential equations are also written using ==. The functional dependence of variables must be indicated explicitly. Derivatives are usually written using the form x'[t] instead of D[x[t],t].

In[136]:= **x''[t] + x[t] == 0**

Out[136]= $x[t] + x''[t] == 0$

We get the solution with DSolve, where the unknown function and the independent variable are given as second and third argument.

In[137]:= **DSolve[x''[t] + x[t] == 0, x[t], t]**

Out[137]= $\{\{x[t] \to C[2]\, \text{Cos}[t] - C[1]\, \text{Sin}[t]\}\}$

The constants `C[1]` and `C[2]` must be determined from the initial conditions. If these are already known, we write the differential equation together with the initial conditions as a set of equations.

In[138]:= **DSolve[{x''[t] + x[t] == 0, x[0] == 1, x'[0] == 0}, x[t], t]**

Out[138]= {{x[t] → Cos[t]}}

As with algebraic equations, we can get the resulting functions as follows:

In[139]:= **x[t] /. %[[1]]**

Out[139]= Cos[t]

Differential equations are also seldom solvable in closed form.

In[140]:= **DSolve[x''[t] + Sin[x[t]] == 0, x, t]**

    Solve::verif :
     Potential solution {x[t] → ComplexInfinity} cannot be verified
       automatically. Verification may require use of limits.

    Solve::ifun : Inverse functions are being
      used by Solve, so some solutions may not be found.

    Solve::verif :
     Potential solution {x[t] → ComplexInfinity} cannot be verified
       automatically. Verification may require use of limits.

    Solve::ifun : Inverse functions are being
      used by Solve, so some solutions may not be found.

    Solve::verif :
     Potential solution {x[t] → ComplexInfinity} cannot be verified
       automatically. Verification may require use of limits.

    General::stop : Further output of
      Solve::verif will be suppressed during this calculation.

    Solve::ifun : Inverse functions are being
      used by Solve, so some solutions may not be found.

    General::stop : Further output of
      Solve::ifun will be suppressed during this calculation.

Out[140]= DSolve[Sin[x[t]] + x″[t] == 0, x, t]

In such cases we have to resort to NDSolve which gives us at least a numerical solution.

In[141]:= **NDSolve[{x''[t] + Sin[x[t]] == 0, x[0] == 1, x'[0] == 0},**
          x[t], {t, 0, 10}]

Out[141]= {{x[t] → InterpolatingFunction[{{0., 10.}}, <>][t]}}

The resulting numerical function can be extracted as usual:

In[142]:= **x[t] /. %[[1]]**

Out[142]= InterpolatingFunction[{{0., 10.}}, <>][t]

and evaluated

In[143]:= **% /. t -> 1.5**

Out[143]= 0.166936

or, as we will see in Part 2, plotted.

In[144]:= **Plot[%%, {t, 0, 10}]**

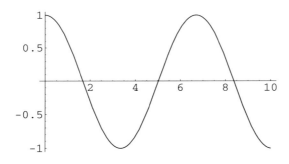

Out[144]= - Graphics -

■ In Depth

● Solutions of Differential Equations as Pure Functions

It is often more practical to request the solution of a differential equation as transformation rule for x itself.

In[145]:= **DSolve[{x''[t] + x[t] == 0, x[0] == 1, x'[0] == 0}, x, t]**

Out[145]= {{x → (Cos[#1] &)}}

This creates a so-called *pure function* (see Section 3.2.3), which can be evaluated exactly as above

In[146]:= **x[t] /. %[[1]]**

Out[146]= Cos[t]

Pure functions help us to verify the solution.

In[147]:= **{x''[t] + x[t] == 0, x[0] == 1, x'[0] == 0} /. %%[[1]]**

Out[147]= {True, True, True}

■ Exercises

● Limits

Calculate the limits of $2^{-\frac{1}{x}}$ as x approaches 0 from the left and from the right.

● Derivatives

Calculate the derivative of x^{x^x}.

Did you derive x^(x^x) or (x^x)^x? Is there a difference?

Calculate the second derivative of $\sin(f(t))\cos(f(t))$ with respect to t, when $f(t)$ is any function of t.

● Integrals

Note the following expression:

Exp[-x] Sin[x]²

Calculate the indefinite integral and the definite integrals in the intervals $[-1, 1]$ and $[0, \infty)$.

Use also the numerical function NIntegrate to determine both definite integrals. How well do the symbolic and numerical results agree?

● Differential Equations

Solve the differential equation system $\{x(t) + x'(t) = y(t), x(t) + y'(t) = 1\}$ and simplify the result. (The documentation of DSolve explains how to solve sets of differential equations.)

Part 2: Graphics

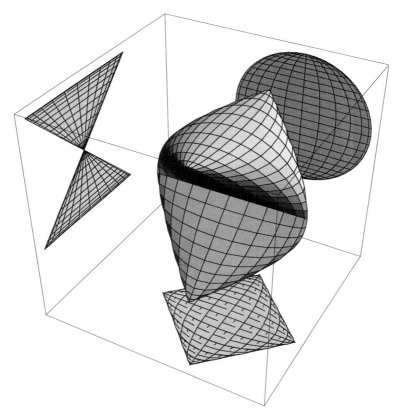

Graphics are an eye-catching, attractive element of *Mathematica*. This part deals with the different methods of producing and refining plots of functions or data and with the export of graphics to other programs.

■ 2.1 Graphs of Functions of One Variable

First a word about the terminology: The term *graph* has a mathematical definition as a set. A computer program can only visualize a finite part of this–possibly infinite–set. Furthermore, we often use axes, headings, etc. to expand on the information. The resulting object will still be called a "graph". *Mathematica* can also create *graphics* which are not graphs. In our terminology a "graph" is a special kind of "graphic".

When visualizing mathematical functions and mappings it is important to consider the dimensions of the domain and of the range. This immediately leads to the appropriate *Mathematica* function. The palette **BasicCalculations > Graphics** contains templates for the common cases.

We start by plotting graphs of functions $\mathbb{R} \to \mathbb{R}$.

First we create the graph of the function $x \to \sin(x)$ over a period using Plot.

In[1]:= **Plot[Sin[x], {x, 0, 2 Pi}]**

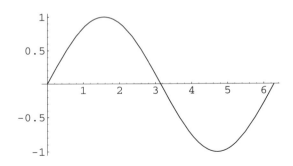

Out[1]= - Graphics -

Since we are mainly interested in the graphics themselves, we will suppress the output cells using a semicolon.

We can also plot several functions at once by passing a list.

In[2]:= **Plot[{Sin[x], Sin[2 x], Sin[3 x]}, {x, 0, 2 π}];**

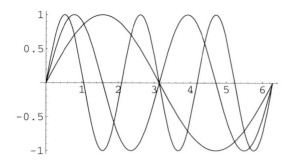

Naturally the function can also first be defined and then plotted.

In[3]:= **function1[x_] = $\dfrac{Sin[x]}{x}$;**

In[4]:= **Plot[function1[x], {x, 0, 2 π}];**

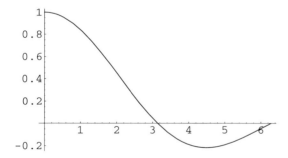

The above definition was written using an immediate definition, since this is just a short-cut and the right-hand side of the definition does not need to be evaluated every time it is called up. A delayed definition produces the same result by evaluating the function definition for every point of the plot. For large calculations, the CPU times of these two variants can differ significantly–but either one can be faster (see the exercise "Efficiency").

In[5]:= **function2[x_] := $\dfrac{Sin[x]}{x}$**

In[6]:= **Plot[function2[x], {x, 0, 2 π}];**

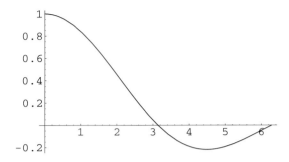

We can change the appearance of graphics in various ways using *options*. The options are set as transformation rules of the form *name* -> *value*. The documentation on Plot in the *Help Browser* lists the specific options of Plot (Compiled etc.). It tells us that all options of Graphics objects can be used as well. In the documentation on Graphics we find a long list of options (from AspectRatio to Ticks) with their default values. Each option again has a documentation of its own.

Let us take a look at some of the most-used options, with which we can edit our graphic. An immediate definition gives it a name for later reference:

In[7]:= **demoPlot = Plot[x^2 Sin[1/x], {x, -π, π}];**

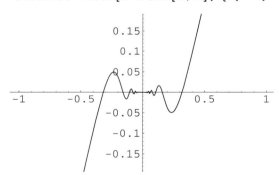

As we can see, the range in this case is automatically limited by *Mathematica*. To clearly display the linear behavior of this function at large arguments, we use PlotRange->All. This plots all values.

In[8]:= **Plot$\left[\mathbf{x}^2 \, \mathbf{Sin}\left[\dfrac{1}{\mathbf{x}}\right], \, \{\mathbf{x}, \, -\pi, \, \pi\}, \, \mathbf{PlotRange} \rightarrow \mathbf{All}\right]$;**

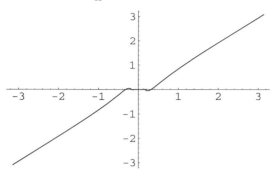

The option value `AspectRatio->Automatic` scales both axes identically. We can add a value for `ImageSize` to get a graphic of a given size.

In[9]:= **Show[%, AspectRatio -> Automatic, ImageSize -> 150];**

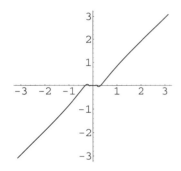

Instead of re-calculating the whole graphic after changing an option, it can also just be re-drawn with the `Show` command and any changed options.

`AxesOrigin` moves origin of the coordinate system.

In[10]:= **Show[demoPlot, AxesOrigin → {-1, 0}];**

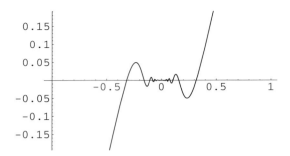

AxesLabel labels the axes using *strings* that must be placed in quotation marks.

In[11]:= **Show[demoPlot, AxesOrigin → {-1, 0},**
        AxesLabel → {"x", "x² Sin[x]"}];

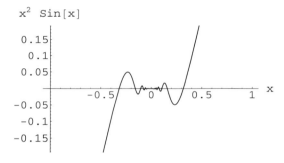

PlotLabel adds a title.

In[12]:= **Show[demoPlot, AxesOrigin → {-1, 0},**
        PlotLabel → "Plot of x² Sin[x]"];

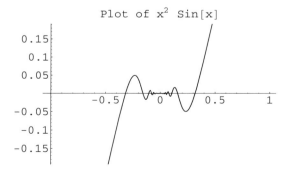

- **In Depth**

• **Frames**

Frame->True creates a frame around the graphic.

In[13]:= **Show[demoPlot, Frame -> True];**

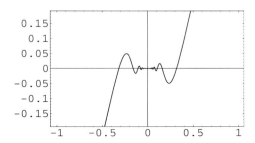

Axes->False suppresses the axes.

In[14]:= **Show[demoPlot, Frame -> True, Axes -> False];**

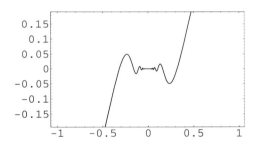

GridLines->Automatic shows a grid.

In[15]:= **Show[demoPlot, Frame -> True, GridLines -> Automatic];**

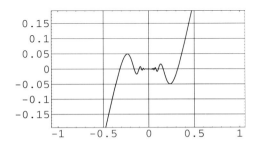

FrameLabel creates a label for the axes (the vertical label appears vertically on the printout).

In[16]:= **Show[demoPlot, Frame -> True, FrameLabel -> {"x", "x² Sin[x]"}];**

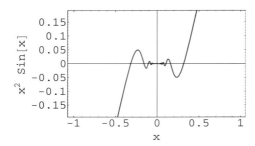

• Changing the Character Format

The option TextStyle uses a list of sub-options with which we can change the format of the text.

In[17]:= **Show[demoPlot, TextStyle ->**
         {FontFamily -> "Times", FontSlant -> "Italic", FontSize -> 9}];

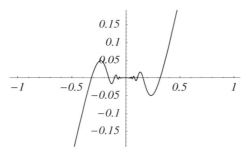

If we want to change only the title:

In[18]:= **Show[demoPlot, AxesOrigin → {-1, 0},**
         PlotLabel → StyleForm["Plot of x² Sin[x]",
           FontFamily -> "Times", FontSlant -> "Italic", FontSize -> 12]];

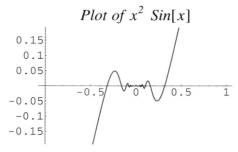

StyleForm lets us use a pre-defined notebook style.

In[19]:= **Show[demoPlot, AxesOrigin → {-1, 0},**
PlotLabel → StyleForm["Plot of x² Sin[x]", "Section"]];

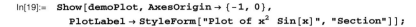

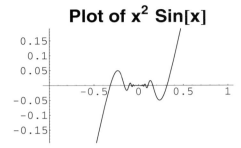

Or we can put a formula into the title.

In[20]:= **Show[demoPlot, AxesOrigin → {-1, 0},**
PlotLabel → TraditionalForm[x² Sin[x]],
TextStyle -> {FontFamily -> "Times", FontSize -> 9}];

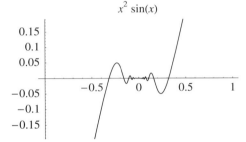

• Changing Lines

The option PlotStyle can be used to change the thickness and color of lines and to draw dashed lines. Because this is an option of Plot and not of Graphics, displaying with Show will not work and the graph must be re-calculated. AbsoluteThickness sets the line thickness to the amount of pixels given.

In[21]:= **Plot$\left[x^2 \; \text{Sin}\left[\dfrac{1}{x}\right], \; \{x, \; -\pi, \; \pi\}, \; \text{PlotStyle} \; \text{-> AbsoluteThickness[2]}\right]$;**

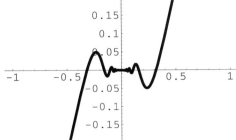

The option value `Dashing` defines dashed lines. Its argument determines the lengths of the segments. As with line thickness, there are two versions: those whose values are given as a fraction of the width of the graphic (`Thickness`, `Dashing`) and those which use an absolute number of pixels (`AbsoluteThickness`, `AbsoluteDashing`).

In[22]:= **Plot$\left[x^2 \text{ Sin}\left[\dfrac{1}{x}\right]\right.$, {x, -$\pi$, π}, PlotStyle -> Dashing[{.1, .02}]];**

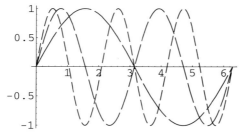

When several graphs are plotted, the functions are handled cyclically by `PlotStyle`.

In[23]:= **Plot[{Sin[x], Sin[2 x], Sin[3 x]}, {x, 0, 2 π}, PlotStyle ->**
          {Dashing[{.2, .02}], Dashing[{.1, .02}], Dashing[{.05, .02}]}];

Colors can easily be defined with the function `Hue`, which represents the color circle at full brightness and saturation when a single argument in the interval [0, 1] is given.

```
In[24]:= Plot[Evaluate[Table[x^n, {n, 0, 10}]],
         {x, 0, 1}, PlotStyle -> Table[Hue[n / 11], {n, 0, 10}],
         PlotRange -> All, AspectRatio -> Automatic];
```

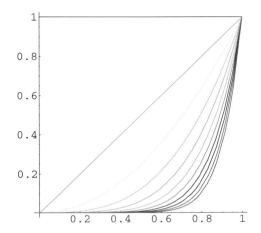

In this calculation we used the functions Table to create a list and Evaluate to enforce the evaluation of the first argument before the Plot function itself. The latter is necessary because Plot only processes lists which are given explicitly.

• Suppressing the Display of Graphics

It is sometimes useful to suppress the display of a graphic. We can do this using the option Display-Function->Identity.

```
In[25]:= Show[demoPlot, DisplayFunction -> Identity]
```

```
Out[25]= - Graphics -
```

DisplayFunction->$DisplayFunction regenerates the display. ($DisplayFunction is a *global variable*.)

```
In[26]:= Show[%, GridLines -> Automatic, DisplayFunction -> $DisplayFunction];
```

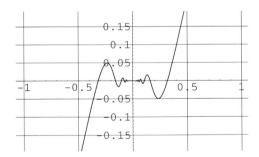

• Poles and Singularities

The default option values of `Plot` produce reasonable plots even when there are poles or singularities.

In[27]:= **Plot[1 / (x - 1), {x, -1, 3}];**

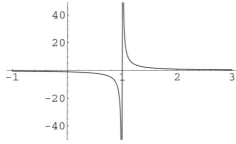

In[28]:= **Plot[Sin[1 / x], {x, 0, .1}];**

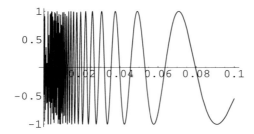

In this case the default value for `PlotPoints` should probably be increased, i.e. the number of function values calculated initially, before the algorithm refines the graph adaptively.

In[29]:= **Plot[Sin[1 / x], {x, 0, .1}, PlotPoints -> 200];**

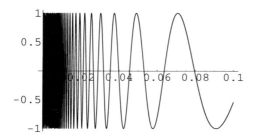

• Superimposing Graphics

`Show` can also be used to superimpose several graphics onto one another.

In[30]:= **Plot[Exp[-x], {x, 0, 2 Pi}];**

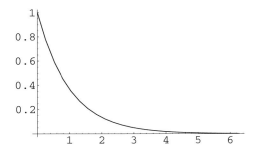

In[31]:= **Plot[Sin[x], {x, 0, 2 Pi}];**

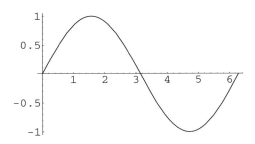

In[32]:= **Plot[Sin[x] Exp[-x], {x, 0, 2 Pi}];**

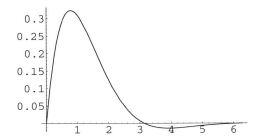

In[33]:= **Show[%, %%, %%%];**

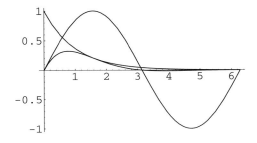

In Section 2.4 we will look at additional helpful tools for designing graphics.

■ Exercises

● Plotting Graphs

Separately plot the graphs of the functions $x \to \sinh(x)$, $x \to \cosh(x)$, and $x \to \tanh(x)$ in the interval $[-2, 2]$.

● Several Graphs at Once

Now draw the above graphs in one figure.

Distinguish between the three curves using various dashed lines and/or colors.

● Variations

Draw a frame with a grid around the above graphic.

Label the x axis and add a title.

Use the *Times* font.

● The Arc Sine

Plot the graph of the function $x \to \arcsin(x)$. (What is the appropriate domain of definition?)

Label both axes.

Use `TraditionalForm` for labeling the ordinate.

● Efficiency

The function `Timing` displays the time used to evaluate an expression. Compare the timings of plots of the following two functions:

```
timingTest1[x_] = Expand[(1 + Sin[x])^50];

timingTest2[x_] := Expand[(1 + Sin[x])^50];
```

Interpret the results.

■ 2.2 Graphs of Functions of Two Variables

In this chapter we look at graphs of functions $\mathbb{R}^2 \to \mathbb{R}$. We can visualize them as surfaces, contour lines, or density plots.

The corresponding *Mathematica* functions are listed in the **BasicCalculations > Graphics** palette or in the *Help Browser*.

■ 2.2.1 Surfaces

In a surface display the rectangular domain appears as the base of a cuboid and the function values are drawn vertically. The graph then is a surface over the base of the cuboid.

In[34]:= **saddle = Plot3D[$x^2 - y^2$, {x, -1, 1}, {y, -1, 1}];**

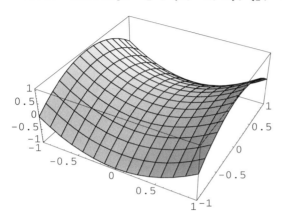

Several of the various options (see the documentation on `Plot3D` and `Graphics3D`) have the same name and work the same way as in two dimensions.

In[35]:= **Show[saddle, PlotLabel → $x^2 - y^2$];**

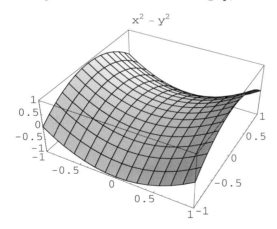

Here and in the in-depth section we will deal only with the most important additional options for three-dimensional objects using examples.

With `ViewPoint` we can change the perspective. A nice tool for this is available from the menu **Input > 3D ViewPoint Selector**. A cube will appear that can be turned with the mouse or by entering the coordinates of the point of view. We move the front edge upwards. After pressing the **Paste** button, the following cell is created:

```
ViewPoint -> {1.306, -3.120, 0.109}
```

This rule can be copied and pasted into the `Show` function. Alternatively we can prepare a cell with the `Show` command, move the insertion mark to the appropriate position, and then use the **3D ViewPoint Selector** to paste directly into `Show`.

In[36]:= **Show[saddle, ViewPoint -> {1.306, -3.120, 0.109}];**

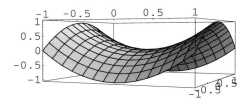

■ In Depth

● Changing the Box and the Axes

The `Boxed` option controls the drawing of the surrounding box.

In[37]:= **Show[saddle, Boxed -> False];**

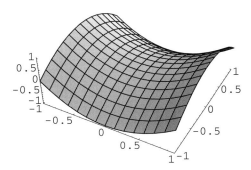

`AxesEdge` changes the positioning of the axes. Enter a list of three pairs which determine for the *x*, *y*, and *z* axes whether they are to be drawn at the side of the bounding box with larger (+1) or smaller (−1) coordinates.

In[38]:= **Show[saddle, AxesEdge -> {{-1, -1}, {-1, 1}, {1, 1}}];**

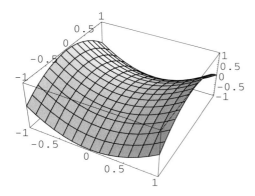

Axes->False hides the axes.

In[39]:= **Show[%%, Axes -> False];**

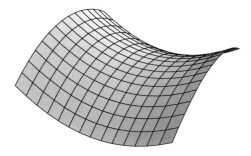

• Colors

The option value Lighting->False turns off the lighting. Polygons are then shaded according to their height.

In[40]:= **Show[%, Lighting -> False];**

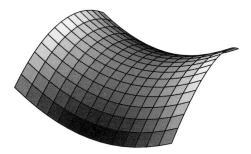

For diffuse ambient lighting we set:

In[41]:= **Show[saddle, AmbientLight -> Hue[1]];**

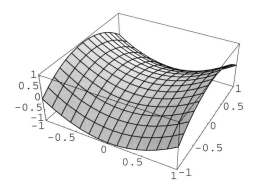

The colors of the lines and the font can be changed using DefaultColor.

In[42]:= **Show[saddle, DefaultColor -> Hue[.6]];**

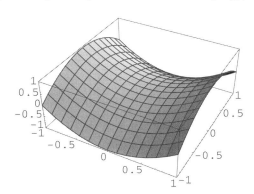

The three `LightSources` have the following default settings:

```
LightSources -> {{{1, 0, 1}, RGBColor[1, 0, 0]},
    {{1, 1, 1}, RGBColor[0, 1, 0]}, {{0, 1, 1}, RGBColor[0, 0, 1]}}
```

`RGBColor` defines a color in the *Red*, *Green*, and *Blue* color model. The preceding lists determine the coordinates of the corresponding light sources.

```
In[43]:= Show[saddle, LightSources -> {{{1, 0, 1}, RGBColor[1, 0, 0]},
    {{1, 1, 1}, RGBColor[0, 1, 0]}, {{0, 1, 1}, RGBColor[0, 0, 1]}}];
```

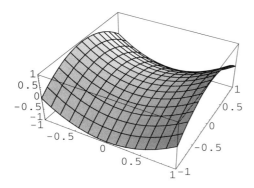

We can change the light sources.

```
In[44]:= Show[saddle, LightSources -> {{{0, -1, 1}, RGBColor[1, 0, 0]},
    {{0, 0, -1}, RGBColor[0, 1, 0]}, {{0, 1, 1}, RGBColor[0, 0, 1]}}];
```

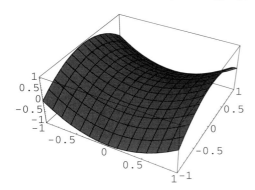

• A Sphere, First Attempt

Let us try to draw a unit sphere.

In[45]:= **Plot3D[Sqrt[1 - x^2 - y^2], {x, -1, 1}, {y, -1, 1}];**

```
Plot3D::gval : Function value 0. + 1. I
   at grid point xi = 1, yi = 1 is not a real number.

Plot3D::gval : Function value 0. + 0.857143 I
   at grid point xi = 1, yi = 2 is not a real number.

Plot3D::gval : Function value 0. + 0.714286 I
   at grid point xi = 1, yi = 3 is not a real number.

General::stop : Further output of
   Plot3D::gval will be suppressed during this calculation.
```

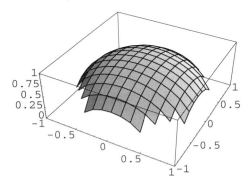

Various problems arise:

• We can only use a rectangle as the domain. Outside the unit circle the root becomes complex. *Mathematica* therefore produces error messages.

• The rectangular grid causes ugly slices.

• Plot3D can display only *one* function, we therefore lose the bottom half of the sphere. This problem could be solved by plotting the bottom half of the sphere separately and then combining the two halves with a Show command.

We will later produce an acceptable figure by a suitable parametric plot.

■ 2.2.2 Contours

Often a different visualization of functions $\mathbb{R}^2 \to \mathbb{R}$ is useful. ContourPlot displays the rectangular domain from above and shows the contours of constant values of the function.

In[46]:= **saddle2 =**
ContourPlot[x² – y², {x, -1, 1}, {y, -1, 1}, ImageSize -> 180];

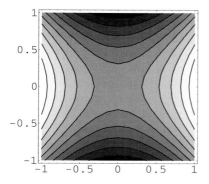

The options belonging to ContourPlot can be looked up in the documentation for ContourPlot and ContourGraphics.

Shading can be drawn in color or left out entirely.

In[47]:= **Show[saddle2, ColorFunction -> Hue, ImageSize -> 180];**

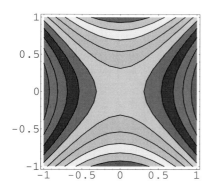

In[48]:= **Show[saddle2, ContourShading -> False, ImageSize -> 180];**

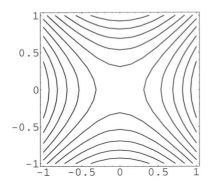

The Contours option controls the contours to be drawn. We can enter either their number or a list of the desired values.

In[49]:= **Show[%, Contours -> 30, ImageSize -> 180];**

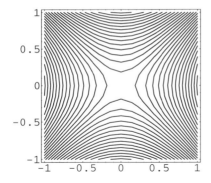

In[50]:= **Show[%, Contours -> {0}, ImageSize -> 180];**

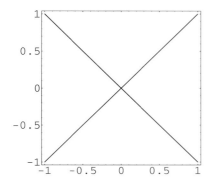

These lines are, of course, the roots of $x^2 - y^2$.

■ 2.2.3 Density Plots

DensityPlot represents the function values on a grey or color scale.

In[51]:= **DensityPlot[x² - y², {x, -1, 1}, {y, -1, 1}, ImageSize -> 180];**

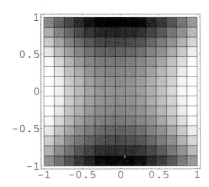

Let us make the grid finer using PlotPoints.

In[52]:= **saddle3 = DensityPlot[x² - y², {x, -1, 1},**
          {y, -1, 1}, PlotPoints -> 50, ImageSize -> 180];

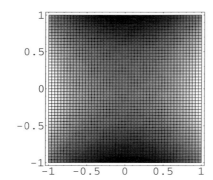

The colored version gives a clearer picture, especially on-screen.

In[53]:= **Show[%, ColorFunction -> Hue, ImageSize -> 180];**

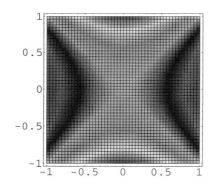

This looks similar to the colorized version of ContourPlot.

■ In Depth

• Converting Graphics

We can interchange the different three-dimensional graphics formats.

In[54]:= **Show[ContourGraphics[%], ImageSize -> 160];**

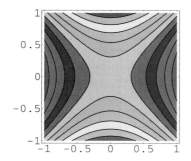

In[55]:= **Show[SurfaceGraphics[%], ImageSize -> 160];**

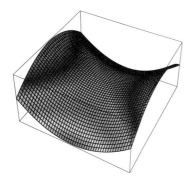

In[56]:= **Show[DensityGraphics[saddle], ImageSize -> 160];**

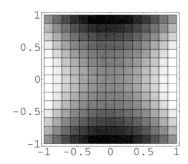

■ Exercises

● Plotting Graphs

Plot the graph of the function $(x, y) \rightarrow \sin(x\,y)$ once as a surface, once with contours, and once as a density plot. Use the rectangle $[0, 2\pi] \times [0, 2\pi]$ as the domain.

Change the value of `PlotPoints` to make the pictures look good.

● Variations

Colorize the contour and density plots.

Turn the surface so that you can see it from below.

● The Arc Tangent

Plot the graph of the function $(x, y) \rightarrow \arctan\left(\frac{y}{x}\right)$ as a surface. Choose the rectangle $[-1, 1] \times [-1, 1]$ as the domain.

The surface is probably not what you expected. For every point in the plane we should be plotting the angle between the x-axis and the line from the origin to the point. Therefore the jump of the surface along the y axis looks a little strange. It has to do with the choice of branch cuts in *Mathematica*'s `ArcTan` function. Study its documentation and find a better solution.

■ 2.3 Parametric Plots

With *parametric plots* we can visualize mappings $\mathbb{R} \rightarrow \mathbb{R}^2$, $\mathbb{R} \rightarrow \mathbb{R}^3$, or $\mathbb{R}^2 \rightarrow \mathbb{R}^3$ by drawing the image of the domain of definition under the mapping. Depending on the dimension of the domain, we get curves or surfaces.

■ 2.3.1 Two-Dimensional Parametric Plots

`ParametricPlot` deals with parametric representations of planar curves, i.e. mappings $\mathbb{R} \rightarrow \mathbb{R}^2$. The x-y coordinates are given as a list.

In[57]:= **ParametricPlot[{Sin[t], Sin[2 t]}, {t, 0, 2 Pi}];**

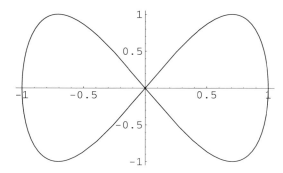

Plotting several curves at once is also possible.

In[58]:= **ParametricPlot[**
   {{Sin[t], Sin[2 t]}, {Sin[t], Sin[4 t]}}, {t, 0, 2 Pi}];

The same options as for `Plot` (and `Graphics`) can also be used here.

■ 2.3.2 Three-Dimensional Parametric Plots

We will now look at mappings $\mathbb{R} \to \mathbb{R}^3$ or $\mathbb{R}^2 \to \mathbb{R}^3$. Their images are curves or surfaces in $\mathbb{R}^3$. `ParametricPlot3D` can be used for both.

First we plot two space curves given by their parametric representations.

In[59]:= **ParametricPlot3D[{Sin[t], Sin[2 t], Sin[4 t] / 2}, {t, 0, 2 Pi}];**

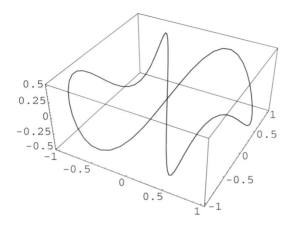

In[60]:= **ParametricPlot3D[{Cos[φ], Sin[φ], φ}, {φ, 0, 4 Pi}];**

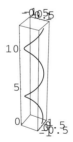

The second figure becomes clearer if we draw a cuboid with equal edge lengths.

In[61]:= **Show[%, BoxRatios -> {1, 1, 1}];**

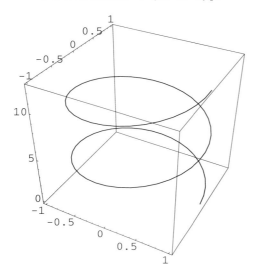

Parametric plots can generate surfaces which do not correspond to graphs of (unique) functions $\mathbb{R}^2 \to \mathbb{R}$. The surface of the unit sphere is such an example, because both signs are possible when the implicit definition $x^2 + y^2 + z^2 = 1$ is solved for a variable.

But we can parametrize the surface of the sphere with spherical coordinates.

In[62]:= **x[θ_, ψ_] = Sin[θ] Cos[ψ];**

In[63]:= **y[θ_, ψ_] = Sin[θ] Sin[ψ];**

In[64]:= **z[θ_] = Cos[θ];**

The surface of the sphere is then the image of the rectangle $[0, \pi] \times [0, 2\pi)$ under the above mapping.

In[65]:= **ParametricPlot3D[**
          {x[ϑ, ψ], y[ϑ, ψ], z[ϑ]}, {ϑ, 0, π}, {ψ, 0, 2 π}];

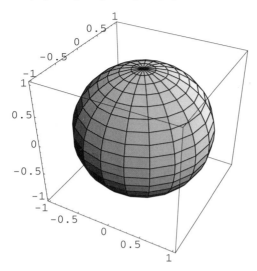

The definitions for x, y, and z are somewhat awkward. These will be used all through the current session whenever the patterns x[..., ...] etc. appear.

In[66]:= **z[1]^2**

Out[66]= $\mathrm{Cos}[1]^2$

It is better to clear the definitions and create the plot directly:

In[67]:= **Clear[x, y, z]**

In[68]:= **ParametricPlot3D[{Sin[θ] Cos[ψ], Sin[θ] Sin[ψ], Cos[θ]},
            {θ, 0, π}, {ψ, 0, 2 π}, ImageSize -> 200];**

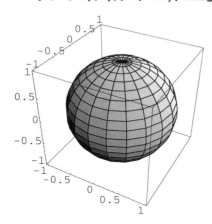

Let us "cut the sphere open"

In[69]:= **ParametricPlot3D[{Sin[θ] Cos[ψ], Sin[θ] Sin[ψ], Cos[θ]},
            {θ, π / 4, π}, {ψ, 0, 2 π}, ImageSize -> 200];**

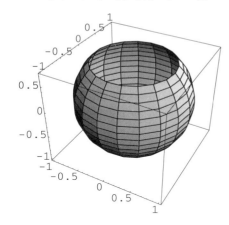

and move closer and upwards to have a better look inside.

In[70]:= **Show[%, ViewPoint -> {0.313, -0.406, 0.859}, ImageSize -> 200];**

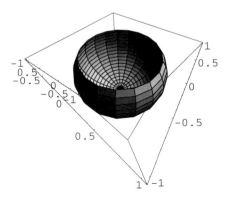

Small changes produce other lovely surfaces:

In[71]:= **ParametricPlot3D[{Sin[ϑ] Cos[ψ], Sin[ϑ] Sin[ψ], Cos[3 ϑ]},**
          {ϑ, 0, π}, {ψ, 0, 3 π / 2}, ImageSize -> 200];

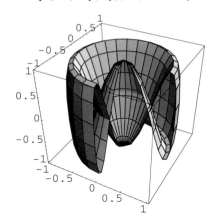

In[72]:=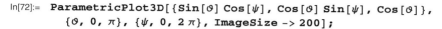

```
ParametricPlot3D[{Sin[ϑ] Cos[ψ], Cos[ϑ] Sin[ψ], Cos[ϑ]},
    {ϑ, 0, π}, {ψ, 0, 2 π}, ImageSize -> 200];
```

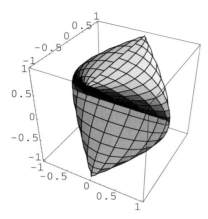

■ Exercises

• A Torus

A torus can be parametrized by the following x, y, z coordinate functions:

$$\{Cos[\varphi] \; (a + b \, Cos[\psi]), \; Sin[\varphi] \; (a + b \, Cos[\psi]), \; b \, Sin[\psi]\}$$

Set $a = 2$ and $b = 1$. Then plot the image of the rectangle $[0, 2\pi) \times [0, 2\pi)$. What is the meaning of the parameters a and b?

Note: if you use transformation rules for inserting values, you may see a message saying that the function to be plotted cannot be compiled. In this case it is better to use Evaluate:

```
PlotFunction [
    Evaluate[{Cos[φ] (a + b Cos[ψ]), …} /. {a → 2, b → 1}],
    {…}, {…}]
```

• Cutting Objects Open

With the selection $a = 1$ and $b = 2$ the above torus surface intersects itself. Convince yourself by "cutting the object open".

■ 2.4 Tools from Standard Packages

Many additional tools are defined in the standard packages which come with *Mathematica* (see *Help Browser* > **Add-ons** > **Standard Packages** > **Graphics**). They can be loaded on demand. A few of them shall be discussed here. (In Version 3.0.x the hyperlinks to functions from the standard packages do not work yet. This will be added in future versions.)

The command

In[73]:= **<< Graphics`**

makes all definitions in the directory Graphics` available. It must be evaluated *before* any of the examples below.

■ 2.4.1 Three-Dimensional Contour Plots

The Graphics`ContourPlot3D` package contains the function ContourPlot3D, which is similar to ContourPlot. It plots the surfaces on which a mapping $\mathbb{R}^3 \to \mathbb{R}$ takes on constant values.

Without explicit settings for the Contours option only the roots will be drawn.

In[74]:= **ContourPlot3D[x^2 - y^2 + z^2, {x, -2, 2},**
         {y, -2, 2}, {z, -2, 2}, ImageSize -> 200];

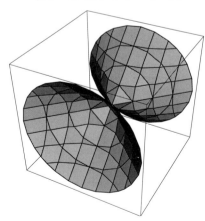

This plots the surfaces corresponding to the function values -1, 0, and 1:

```
In[75]:= ContourPlot3D[x^2 - y^2 + z^2, {x, -2, 2}, {y, -2, 2},
         {z, -2, 2}, Contours -> {-1., 0., 1.}, ImageSize -> 200];
```

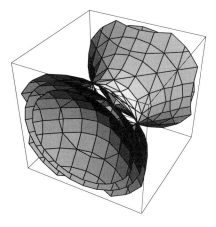

■ 2.4.2 Tools for Two-Dimensional Graphics

The `Graphics`Graphics`` package contains various tools for logarithmic plots, bar and pie charts, data plots with error bars, etc. The documentation is worth a look. We will restrict ourselves to two examples.

```
In[76]:= LogPlot[Cosh[x], {x, 0, 10}];
```

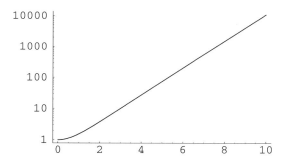

In[77]:= **BarChart[{1, 3, 5, 3, 1}];**

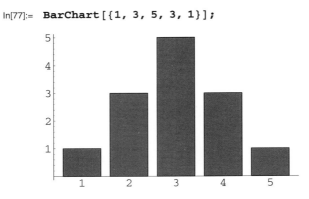

■ 2.4.3 Tools for Three-Dimensional Graphics

The Graphics`Graphics3D` package is also worth a look.

Let us use ShadowPlot3D and Shadow to project surfaces onto the bounding box. This can be very instructive.

In[78]:= **ShadowPlot3D[Sin[x - y],**
{x, 0, Pi}, {y, 0, Pi}, ImageSize -> 200];

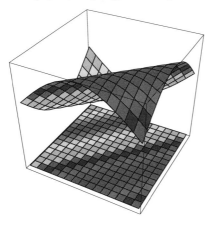

In[79]:= **ParametricPlot3D[**
{Sin[ϑ] Cos[ψ], Cos[ϑ] Sin[ψ], Cos[ϑ]}, {ϑ, 0, π},
{ψ, 0, 2 π}, PlotPoints -> {25, 25}, ImageSize -> 200];

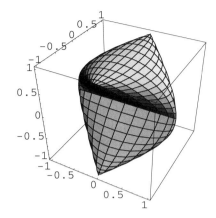

In[80]:= **Shadow[%, ImageSize -> 200];**

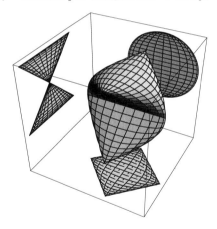

■ 2.4.4 Legends

The legends defined in Graphics`Legend` can be changed by many options (see documentation).

In[81]:= `Plot[{Sin[x], Cos[x]}, {x, -2 π, 2 π},`
`        PlotStyle → {GrayLevel[0], Dashing[{.03}]},`
`        PlotLegend → {"Sine", "Cosine"}];`

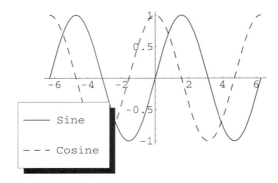

In[82]:= `DensityPlot[Sin[x² + y²], {x, -3, 3}, {y, -3, 3},`
`          PlotPoints -> 50, ColorFunction -> Hue, ImageSize -> 180];`

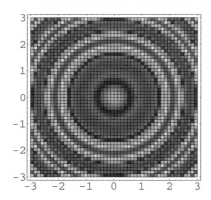

The range is the interval [–1, 1]. This is mapped onto [0, 1] by `ColorFunction`. We must therefore label the legend with values between –1 and 1.

In[83]:= **ShowLegend[%, {Hue, 10, "-1", "1"}];**

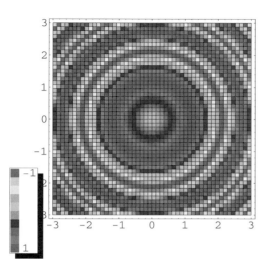

■ 2.4.5 Vector Fields

A useful visualization of mappings $\mathbb{R}^2 \to \mathbb{R}^2$ or $\mathbb{R}^3 \to \mathbb{R}^3$ can be obtained by drawing the image vector as an arrow in each grid point of the domain. The corresponding documentation can be found in the `Graphics`PlotField`` and `Graphics`PlotField3D`` packages.

The following vector field belongs to a mathematical pendulum:

In[84]:= **PlotVectorField[{y, -Sin[x]},**
      {x, -Pi, 2 Pi}, {y, -Pi, Pi}, Axes -> True];

With the three-dimensional function `PlotVectorField3D` the velocity field of a rotation around the *z* axis can be viewed. The `VectorHeads->True` option setting makes sure that the arrow heads are drawn.

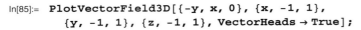

```
In[85]:= PlotVectorField3D[{-y, x, 0}, {x, -1, 1},
           {y, -1, 1}, {z, -1, 1}, VectorHeads → True];
```

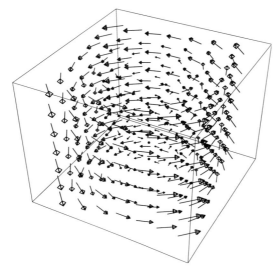

■ In Depth

• Collisions of Names

This paragraph deals with a problem which arises when function names are used from packages which have not yet been loaded (see *Help Browser* > **Add-ons** > **Working with Add-ons** > **Loading Packages**).

To get a clear starting point we first end the active kernel session with **Kernel** > **Quit Kernel**.

Now we attempt to create a logarithmic plot using the following command:

```
In[1]:= LogPlot[Exp[3 x], {x, 0, 2}]
```

```
Out[1]= LogPlot[E³ˣ, {x, 0, 2}]
```

Because the definition of `LogPlot` has not yet been loaded with `<<Graphics`` or `<<Graphics`Graphics``, nothing happens. Therefore we try:

```
In[2]:= << Graphics`
```

But it still does not work:

In[3]:= **LogPlot[Exp[3 x], {x, 0, 2}]**

Out[3]= LogPlot[E$^{3\,x}$, {x, 0, 2}]

The following command shows that LogPlot is now defined in the *global context*, but the name should be in the context of the package. (Further information to contexts can be found in Section 4.4.5.)

In[4]:= **? LogPlot**

Global`LogPlot

We can solve the problem using Remove.

In[5]:= **Remove[LogPlot]**

Now the context is correct and the command works.

In[6]:= **? LogPlot**

Graphics`Graphics`LogPlot

Attributes[LogPlot] = {Stub}

LogPlot = "Graphics.m"

In[7]:= **LogPlot[Exp[3 x], {x, 0, 2}];**

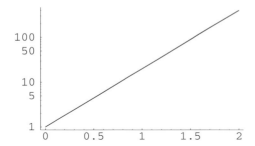

• **Further Parametric Plots**

The Graphics`ParametricPlot3D` package contains, for historical reasons, the ParametricPlot3D function, which we have already seen. Additional useful tools are Spherical-Plot3D and CylindricalPlot3D.

With SphericalPlot3D we need to indicate the radius as function of the spherical coordinate angles ϑ and ψ.

In[8]:= **SphericalPlot3D[1, {ϑ, 0, π}, {ψ, 0, 2 π}];**

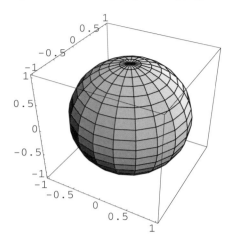

CylindricalPlot3D draws the z coordinate in function of ρ and φ (cylindrical coordinates).

In[9]:= **CylindricalPlot3D[ρ^2, {ρ, 0, 1}, {φ, 0, 2 π}];**

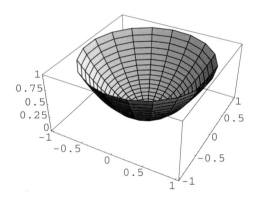

• An Undocumented Tool

The following option value for automated tick marks at multiples of π is unfortunately undocumented. PiScale can be found in the Graphics`Graphics` package.

In[10]:= `Plot[Sin[x], {x, 0, 2 Pi}, Ticks -> {PiScale, Automatic}];`

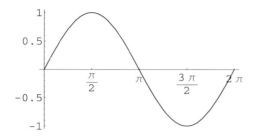

■ Exercises

• A Sphere

Plot the surface of a unit sphere using `ContourPlot3D`.

• Logarithmic Plots

Draw and interpret a logarithmic and a double logarithmic plot of $x \to x^3$.

• Pie Charts

Draw a pie with piece sizes of 1/2, 1/4, 1/6, 1/12.

• Projections of Surfaces

Draw a parametric plot of $\{ \texttt{Sin[} \vartheta \texttt{]Cos[} \psi \texttt{/2], Sin[} \vartheta \texttt{]Sin[} \psi \texttt{], Cos[} \vartheta \texttt{]} \}$ for the parameter domain $\{ \vartheta, 0, \pi \}, \{ \psi, 0, 4\pi \}$.

Study the surface by projecting it onto three planes of the bounding box.

• Legends

Go back to the above example for legends and move the legend to the right-hand side of the plot.

• Cones

Use `CylindricalPlot3D` to draw a cone.

Deform it into a corrugated surface by adding a small φ-dependent sine modulation.

• Vector Fields

Draw a vector field of the mapping $\mathbb{R}^2 \to \mathbb{R}^2$, given by $(x, y) \to (x - y, x + y)$.

■ 2.5 Animations

Sometimes an additional dimension of a problem can be visualized by mapping a parameter (or a variable) on the time. The **Cell > Animate Selected Graphics** command animates a selected group with graphics cells. The cell group can either be created using the functions in the Graphics`Animation` standard package, or "by hand". We will study the second option in the third part, two examples of the first one are shown here. Naturally the animation works only on-screen. The book shows the first figure of the sequence.

If the group of Graphics`packages has not yet been loaded you must at least load the animation package at this point.

In[11]:= **<< Graphics`Animation`**

We create the graphics by varying a parameter with the help of Animate. The list {n, -.4, 1, .2} contains the variable, the initial value, the end value, and finally the step size.

In[12]:= **Animate[Plot[x^4 - n x^2, {x, -1, 1},
          PlotRange → {All, {-.25, 1}}], {n, -.4, 1, .2}]**

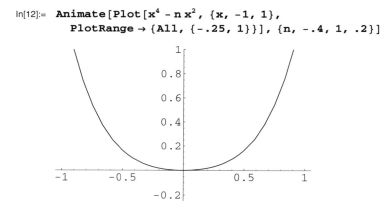

The above cell group can be animated on-screen using the **Cell > Animate Selected Graphics** command or by double-clicking on the graphic. In the bottom left-hand corner of the notebook window a "control board" appears which you can use to change, among other things, the direction and the speed of the animation.

To make the animation run smoothly the axes ranges must be identical for all figures. This is achieved by an explicit setting for PlotRange.

In this example a three-dimensional view would of course also be possible.

In[13]:= **Plot3D[x^4 - n x^2, {x, -1, 1}, {n, -.4, 1}];**

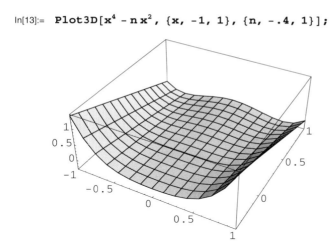

With three-dimensional graphics this is not so easy. You will find an example in the exercises.

Let us draw the "pillow" from above again, this time without box or axes.

In[14]:= **spinDemo = ParametricPlot3D[**
{Sin[ϑ] Cos[ψ], Cos[ϑ] Sin[ψ], Cos[ϑ]}, {ϑ, 0, π},
{ψ, 0, 2 π}, Axes -> False, Boxed -> False, ImageSize -> 180];

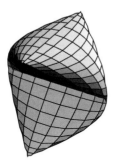

SpinShow rotates the object. Because of the symmetry a half turn is enough.

In[15]:= **SpinShow[spinDemo, Frames -> 10, SpinRange -> {0, Pi}]**

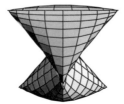

(Close the cell group in the notebook, select it, then choose **Cell > Animate Selected Graphics**.)

Further functions and options can be found in the documentation of Graphics`Animation`.

■ Exercises

● Parameters in a Function of One Variable

Create an animation where the sine function over a period is moved to the right in ten steps. A possible solution can be animated in the notebook:

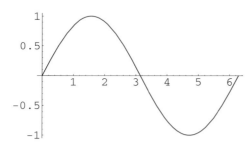

Try out the different buttons on the "control board" in the bottom left corner of the window.

● Parameters in a Function of Two Variables

Look at the function $(x, y) \rightarrow n^2 (\sin x + \sin y)^2 + \cos x + \cos y$ with the parameter n.

First plot the graph in the domain $[-\pi, \pi] \times [-\pi, \pi]$ and for $n = 0$.

Visualize the changes in the surface as the parameter *n* varies in the interval from 0 to 1 (increment length 1/10). Make sure that the "movie" does not show any tears due to incompatible axis scales.

- **Your Own Example**

Construct your own example for a `MovieParametricPlot`.

■ 2.6 Exporting to Other Programs

Not all readers do all their work in *Mathematica*—even though the text system is adequate for many applications, e.g., for writing this book. There might still be some need to export graphics and formulas to other programs.

Exporting formulas is somewhat unsatisfactory, since they lose their mathematical content during export and exist only as graphics. But the technology is exactly the same as for graphics.

The most versatile technique is saving the graphic (or formula) in a file and then importing this file into a word processing or graphic program. The EPS format gives the best results, because it prints perfectly and because the format is available on all platforms. The following steps do the job:

1. Select the graphic.
2. Menu **Edit > Save Selection As... > EPS**.
3. Name the file.
4. **Save** the file.

The *Adobe Illustrator* format can be very useful for owners of the program. It allows the manipulation of the figures in all ways possible in *Illustrator*.

Depending on the computer platform, the menu **Edit > Save Selection As...** may contain other formats. They sometimes give less perfect results after export to files or via the clipboard (**Edit > Copy As**).

■ Exercise

- **Saving a Graphic in a File**

Plot the function $x \to 2^x$ in the interval $[-2, 2]$.

Export the graphic in EPS format and import it into your word processing program.

Part 3: Lists and Graphics Programming

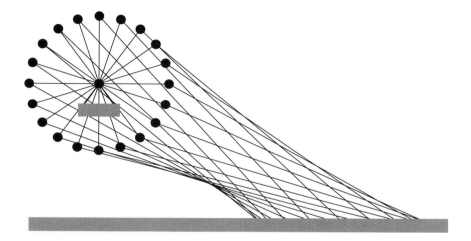

Lists are the most important objects in *Mathematica*. They appear everywhere, visibly or hidden. Once you learn to use them properly, it will make working with the program that much simpler.

In this part we will increase our knowledge of lists in-depth, then apply them to solve simple linear algebra problems and to create our own graphics.

■ 3.1 Lists

■ 3.1.1 Creating One-Dimensional Lists

We already know that lists are objects whose elements are placed into curly braces.

In[1]:= **{1, 4, 9}**

Out[1]= {1, 4, 9}

The *Mathematica* function `Table` is useful for creating lists. It evaluates an expression for different values of an iterator.

In[2]:= **Table[i^2, {i, 10}]**

Out[2]= {1, 4, 9, 16, 25, 36, 49, 64, 81, 100}

The list in the second argument of `Table` contains the name of the iterator and further information. The following forms are possible (i is the iterator):
- {n} creates n identical entries (with no iterator present)
- {i,n} varies i over 1, 2, ..., n
- {i,a,n} varies i over a, a+1, ..., n
- {i,a,n,s} varies i over a, a+s, a+2s, ..., n^*, where $n^* \leqslant$ n

In[3]:= **Table[a, {10}]**

Out[3]= {a, a, a, a, a, a, a, a, a, a}

In[4]:= **Table[i^2, {i, 0, 10}]**

Out[4]= {0, 1, 4, 9, 16, 25, 36, 49, 64, 81, 100}

In[5]:= **Table[i^2, {i, 0, 1, .3}]**

Out[5]= {0, 0.09, 0.36, 0.81}

Arithmetic sequences can be created more easily using `Range`. No name is needed for the iterator in this case.

In[6]:= **Range[10]**

Out[6]= {1, 2, 3, 4, 5, 6, 7, 8, 9, 10}

In[7]:= **Range[0, 10]**

Out[7]= {0, 1, 2, 3, 4, 5, 6, 7, 8, 9, 10}

In[8]:= **Range[0, 1, .3]**

Out[8]= {0, 0.3, 0.6, 0.9}

■ 3.1.2 Manipulating Lists

Many functions are available for the manipulation of lists (see Section 1.8 of the *Mathematica Book*). A few useful examples follow.

Our exercise object is:

In[9]:= **list1 = {a, b, c, d, e}**

Out[9]= {a, b, c, d, e}

The function Length determines its length (what a surprise!).

In[10]:= **Length[list1]**

Out[10]= 5

We already know how to extract elements.

In[11]:= **list1[[2]]**

Out[11]= b

Or in StandardForm:

In[12]:= **list1〚2〛**

Out[12]= b

Take can extract entire sublists. A positive number as second argument results in the same number of elements from the left. Negative numbers count the elements from the right. If we pass a list with a start and end number in the second argument, Take returns the corresponding sublist.

In[13]:= **Take[list1, 3]**

Out[13]= {a, b, c}

In[14]:= **Take[list1, -3]**

Out[14]= {c, d, e}

In[15]:= **Take[list1, {2, 4}]**

Out[15]= {b, c, d}

Drop functions analogously and throws the corresponding elements away.

In[16]:= **Drop[list1, 3]**

Out[16]= {d, e}

In[17]:= **Drop[list1, -3]**

Out[17]= {a, b}

In[18]:= **Drop[list1, {2, 4}]**

Out[18]= {a, e}

RotateRight permutes the elements of a list cyclically to the right.

In[19]:= **RotateRight[list1]**

Out[19]= {e, a, b, c, d}

In[20]:= **RotateRight[list1, 2]**

Out[20]= {d, e, a, b, c}

Sort rearranges the elements according to a sort order.

In[21]:= **Sort[%]**

Out[21]= {a, b, c, d, e}

Lists can be joined using Join.

In[22]:= **Join[list1, {f, g}]**

Out[22]= {a, b, c, d, e, f, g}

Or we can use the function Flatten, which "flattens out" a nested list into a one-dimensional one.

In[23]:= **Flatten[{list1, {f, g}}]**

Out[23]= {a, b, c, d, e, f, g}

Or we can use Append twice, to add an element to the end each time.

In[24]:= **Append[Append[list1, f], g]**

Out[24]= {a, b, c, d, e, f, g}

Position gives the position(s) of a pattern; the corresponding elements can be extracted with Extract.

In[25]:= **list2 = {a^2, b, b, c, d^2, b, e}**

Out[25]= {a^2, b, b, c, d^2, b, e}

In[26]:= **Position[list2, b]**

Out[26]= {{2}, {3}, {6}}

In[27]:= **Extract[list2, %]**

Out[27]= {b, b, b}

In[28]:= **Position[list2, _^2]**

Out[28]= {{1}, {5}}

In[29]:= **Extract[list2, %]**

Out[29]= {a^2, d^2}

Select tests the elements of a list for properties. The list must be given together with a function, which returns True for the desired elements.

In[30]:= **myTest[x_] = (x > 10) && (x < 50)**

Out[30]= x > 10 && x < 50

In[31]:= **Select[Table[i^2, {i, 10}], myTest]**

Out[31]= {16, 25, 36, 49}

One of the exercises shows that lists can also be used for set calculations.

■ 3.1.3 Multidimensional Lists

`Table` can also create multidimensional lists:

In[32]:= **Table[a^i + b^j, {i, 3}, {j, 3}]**

Out[32]= $\{\{a + b, a + b^2, a + b^3\},$
$\{a^2 + b, a^2 + b^2, a^2 + b^3\}, \{a^3 + b, a^3 + b^2, a^3 + b^3\}\}$

The function `MatrixForm` displays the list as a matrix.

In[33]:= **MatrixForm[%]**

Out[33]//MatrixForm=
$$\begin{pmatrix} a + b & a + b^2 & a + b^3 \\ a^2 + b & a^2 + b^2 & a^2 + b^3 \\ a^3 + b & a^3 + b^2 & a^3 + b^3 \end{pmatrix}$$

■ Exercises

• Prime Numbers

Create a list of all odd numbers between 10^6 and $10^6 + 10^3$.

The function `PrimeQ` tests whether a number is prime. Use it and `Select` to determine the prime numbers in the above list.

• Sets

Use the functions `Union` and `Intersection` to form the union and the intersection of the sets $\{a, a, b, c, d, e, f\}$ and $\{a, f, g, g, j\}$. Note that `Union` can also be used to get rid of duplicated elements in a list.

■ 3.2 Calculating with Lists

■ 3.2.1 Automatic Operations

Many functions of one argument are automatically mapped onto the elements of lists.

In[34]:= **Sin[{1, 2, 3}]**

Out[34]= $\{Sin[1], Sin[2], Sin[3]\}$

This multiplies each element of the list by 3:

In[35]:= **3 {a, b, c}**

Out[35]= {3 a, 3 b, 3 c}

The (normal) product is calculated element-wise.

In[36]:= **{a, b, c} {1, 2, 3}**

Out[36]= {a, 2 b, 3 c}

For more involved problems there are also the functions Inner and Outer, which calculate generalized inner and outer products.

Scalar and cross products are predefined in the functions Dot and Cross, even though you could also easily do this yourself (there is an exercise for it).

In[37]:= **Dot[{x, y, z}, {u, v, w}]**

Out[37]= u x + v y + w z

In[38]:= **Cross[{x, y, z}, {u, v, w}]**

Out[38]= {w y - v z, -w x + u z, v x - u y}

We can even write these in the following notation:

In[39]:= **{x, y, z}.{u, v, w}**

Out[39]= u x + v y + w z

In[40]:= **{x, y, z} × {u, v, w}**

Out[40]= {w y - v z, -w x + u z, v x - u y}

■ 3.2.2 Mapping Functions on Lists

The automatic mapping of functions on lists will not solve all our problems. For example, let us look at Variables which delivers the variables in a polynomial.

In[41]:= **Variables[x + y]**

Out[41]= {x, y}

If we apply this function to a list, we get a flat list of all variables in the original list.

In[42]:= **Variables[{x + y, x + z, y}]**

Out[42]= {x, y, z}

But what are the variables of the single elements of the list? An iteration over the elements is somewhat awkward:

In[43]:= **Table[Variables[{x + y, x + z, y}[[i]]], {i, 3}]**

Out[43]= {{x, y}, {x, z}, {y}}

As part of a complicated program, this would have the disadvantage that we first have to determine the length (possible with Length) and that the construction is hard to read. Therefore *Mathematica* offers the possibility of mapping functions of one argument onto lists using Map.

In[44]:= **Map[Variables, {x + y, x + z, y}]**

Out[44]= {{x, y}, {x, z}, {y}}

Because Map appears often in programs, there is also a short notation:

In[45]:= **Variables /@ {x + y, x + z, y}**

Out[45]= {{x, y}, {x, z}, {y}}

■ 3.2.3 Pure Functions

In the example for Select we defined an auxiliary function

In[46]:= **myTest[x_] = (x > 10) && (x < 50);**

to recognize the desired list elements. Often we need this kind of tool only once, so there is no point in giving it a name. This can be avoided with the help of *pure functions*. The pure function for this problem can be written with Function and looks like this:

In[47]:= **Function[x, (x > 10) && (x < 50)]**

Out[47]= Function[x, x > 10 && x < 50]

The first argument of Function contains the name of the auxiliary (local) variable x, the second argument is the expression in x to be evaluated. Such objects are applied in the usual way. In the following example the argument 9 is substituted for x and the function body evaluated.

In[48]:= **Function[x, (x > 10) && (x < 50)][9]**

Out[48]= False

Pure functions can easily be mapped on lists

In[49]:= **Function[x, (x > 10) && (x < 50)] /@ {1, 20, 100, 30}**

Out[49]= {False, True, False, True}

or used in functions like Select.

In[50]:= **Select[Table[i^2, {i, 10}], Function[x, (x > 10) && (x < 50)]]**

Out[50]= {16, 25, 36, 49}

Now we do not need the name x in the pure function either. We can replace it with # and write only the function definition.

In[51]:= **Select[Table[i^2, {i, 10}], Function[(# > 10) && (# < 50)]]**

Out[51]= {16, 25, 36, 49}

Because this also appears often, there is an even shorter way to write it: leave out the Function and delimitate the pure function with a &.

In[52]:= **Select[Table[i^2, {i, 10}], (# > 10) && (# < 50) &]**

Out[52]= {16, 25, 36, 49}

In this way we can easily calculate the partial derivatives of the expression

$$\frac{x - y}{\sqrt{x^2 + y^2 + z^2}};$$

with respect to x, y, z, i.e. the gradient.

In[53]:= $\mathbf{D}\left[\dfrac{\mathbf{x - y}}{\sqrt{\mathbf{x^2 + y^2 + z^2}}}, \text{\#}\right]$ **& /@ {x, y, z}**

Out[53]= $\left\{-\dfrac{x\,(x - y)}{(x^2 + y^2 + z^2)^{3/2}} + \dfrac{1}{\sqrt{x^2 + y^2 + z^2}}, \right.$

$\left. -\dfrac{(x - y)\,y}{(x^2 + y^2 + z^2)^{3/2}} - \dfrac{1}{\sqrt{x^2 + y^2 + z^2}}, -\dfrac{(x - y)\,z}{(x^2 + y^2 + z^2)^{3/2}}\right\}$

In a pure function with several arguments, these are either combined into a list (when Function is used with two arguments) or indicated with #1, #2, ... (in the short notation).

In[54]:= **Function[{x, y, z}, Sqrt[x^2 + y^2 + z^2]]**

Out[54]= Function$\left[\{x, y, z\}, \sqrt{x^2 + y^2 + z^2} \right]$

In[55]:= **%[a, b, c]**

Out[55]= $\sqrt{a^2 + b^2 + c^2}$

In[56]:= **Function[Sqrt[#1^2 + #2^2 + #3^2]][a, b, c]**

Out[56]= $\sqrt{a^2 + b^2 + c^2}$

Of course we can also write definitions with pure functions.

In[57]:= **geometricMean = (#1 #2 #3) ^ (1 / 3) &;**

In[58]:= **geometricMean[a, b, c]**

Out[58]= $(a\,b\,c)^{1/3}$

The alternative definition of the form

In[59]:= **geometricMean1[x_, y_, z_] := (x y z) ^ (1 / 3)**

In[60]:= **geometricMean1[a, b, c]**

Out[60]= $(a\,b\,c)^{1/3}$

is a definition for the pattern geometricMean1[x_,y_,z_]. In contrast, the pure function

In[61]:= **geometricMean = (#1 #2 #3) ^ (1 / 3) &;**

can be interpreted as a definition for the *head* geometricMean (see Section 4.1).

■ 3.2.4 Using List Elements as Arguments

To conclude this first look at some exotic constructions, let us discuss Apply. It is used to apply the elements of a list as arguments of a function.

```
In[62]:= Apply[f, {a, b, c}]
```

```
Out[62]= f[a, b, c]
```

This is, of course, different from:

```
In[63]:= f[{a, b, c}]
```

```
Out[63]= f[{a, b, c}]
```

We now consider the addition function `Plus` which we normally write using the operator + (see Section 4.1). If we apply it to a list using `Apply`, we get the sum of the elements

```
In[64]:= Apply[Plus, {a, b, c}]
```

```
Out[64]= a + b + c
```

or, analogously, the product using `Times`:

```
In[65]:= Apply[Times, {a, b, c}]
```

```
Out[65]= a b c
```

There is also an infix notation:

```
In[66]:= Plus @@ {a, b, c}
```

```
Out[66]= a + b + c
```

```
In[67]:= Times @@ {a, b, c}
```

```
Out[67]= a b c
```

The above function for the geometric mean has the disadvantage that the number of arguments is set at three. We can now define a variation which handles a list of indeterminate length.

```
In[68]:= geometricMean2 = (Times @@ #) ^ (1 / Length[#]) &;
```

```
In[69]:= geometricMean2[{a, b, c, d, e}]
```

$$Out[69]= (a\ b\ c\ d\ e)^{1/5}$$

Of course, we also have built-in functions for indexed sums and products: `Sum` and `Product`. They work like `Table`.

In[70]:= **Sum[a^i, {i, 5}]**

Out[70]= $a + a^2 + a^3 + a^4 + a^5$

■ 3.2.5 Plotting Lists

There are variations to almost all graphic functions to plot lists of values. Their names always begin with List (ListPlot, ListPlot3D, ListContourPlot, etc.). The data can be taken from *Mathematica* itself or can be imported from other programs using ReadList (see the in-depth section).

We create a list with a couple of numerical values of the cosine function.

In[71]:= **cosList = Table[N[Cos[x]], {x, 0, 2 Pi, 2 Pi / 50}];**

ListPlot plots these points.

In[72]:= **ListPlot[cosList];**

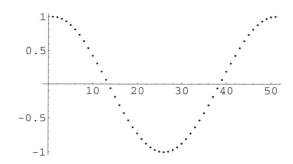

They are joined with the option PlotJoined->True.

In[73]:= **ListPlot[cosList, PlotJoined -> True];**

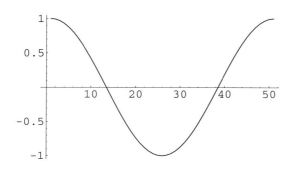

The abscissa is labeled with the entry numbers in the list (1, ... , 51) in this case. But we can also plot a list of point pairs, which gives us useful scales on both axes.

In[74]:= **Short[xCosxList = Table[N[{x, Cos[x]}], {x, 0, 2 Pi, 2 Pi / 50}]]**

Out[74]//Short=
        {{0, 1.}, ≪49≫, {6.28319, 1.}}

In[75]:= **ListPlot[xCosxList];**

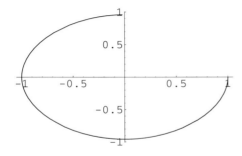

■ In Depth

• Solutions of Differential Equations as Pure Functions

We have seen in the in-depth to Section 1.4.6 how the solution of a differential equation can be requested as a pure function. This form is useful if the derivative of the solution must be calculated (we will look at the detailed reason for it in the in-depth to Section 4.2). It allows us, for instance, to draw a parametric plot in the phase space $\{x(t), x'(t)\}$.

As an example, let us look at the numerical solution of the nonlinear oscillation equation $x''(t) + \sin(x(t)) = 0$ with $\{x(0) = 1, x'(0) = 0\}$.

In[76]:= **NDSolve[{x''[t] + Sin[x[t]] == 0, x[0] == 1, x'[0] == 0}, x, {t, 0, 5}];**

In[77]:= **ParametricPlot[Evaluate[{x[t], x'[t]} /. %[[1]]], {t, 0, 5}];**

(Evaluate makes sure that the transformation rule is applied before the curve is plotted.)

• The Efficiency of Numerical Sums

You will calculate many numerical sums in the exercises. To increase the numerical efficiency, the following consideration is useful.

We first look at the sum:

In[78]:= **Sum[1 / i ^ 3, {i, 10}]**

Out[78]= $\dfrac{19164113947}{16003008000}$

Mathematica calculates a priori with exact rational numbers. Of course this becomes very complicated for larger sums, therefore it is essential to change to approximate numbers as quickly as possible.

In[79]:= **Sum[1 / N[i] ^ 3, {i, 10}]**

Out[79]= 1.19753

After increasing the number of terms, we can look at the difference in CPU times using the function Timing. In addition to the result, it also shows the processing time. The result of the first calculation produces a large fraction, which we will suppress.

In[80]:= **Timing[Sum[1 / i ^ 3, {i, 1000}];]**

Out[80]= {1.68333 Second, Null}

In[81]:= **Timing[Sum[1 / N[i] ^ 3, {i, 1000}];]**

Out[81]= {0.0666667 Second, Null}

• Saving and Reading Lists

We will end this in-depth section with an example for reading external data. First we create a list to work on, one which is comprised of values of the sine function with a superimposed "noise" (created with Random).

In[82]:= **data = Table[Sin[x] + Random[Real, {-.03, .03}], {x, 0, 2 Pi, 2 Pi / 200}];**

In[83]:= **ListPlot[data];**

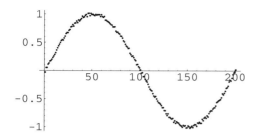

We save the list in a file with the `Save` command.

In[84]:= **Save["data.m", data]**

Now we clear the definition.

In[85]:= **Clear[data]**

In[86]:= **data**

Out[86]= data

The following command reads it again:

In[87]:= **<< data.m;**

In[88]:= **ListPlot[data];**

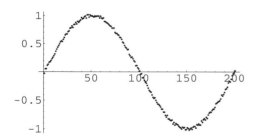

But a look at the file `data.m` shows that the data was saved in the usual *Mathematica* format. It cannot therefore be used as an example for data which was created in another program. Without further explanation, we will accept that the following command writes the data unformatted into single lines:

In[89]:= **PutAppend[#, "pure-data"] & /@ data;**

A look at the file `pure-data` proves it. For this we use either the *Mathematica* command:

!! pure-data

which prints the file onscreen (we will not show this uninteresting printout here), or we could as well open the file using the command **File > Open** in the front end.

Now we can use the function ReadList to read the data from this file again.

In[90]:= **externalData = ReadList["pure-data"];**

In[91]:= **ListPlot[externalData];**

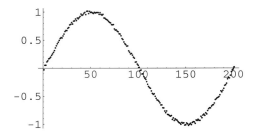

When reading structured data, for example *Excel* sheets which have been saved as text, the sequence of data types can be entered as second argument in ReadList.

■ Exercises

● A Logarithm Table

Create in at least two different ways a table of natural logarithms of the numbers between 1 and 10.

Now create a list which contains the pairs {n, log(n)}. (TableForm displays it as a table.)

How can the above be solved with a pure function which is mapped onto Range[1,10]?

● The Gradient

Define a function named gradient, which calculates the gradient of an expression in the simplest way. The first argument should be the expression, the second a list of the variables.

● Pure Functions

Create with

aList = Table[i + j, {i, 10}, {j, Random[Integer, {1, 10}]}]

a nested list with sublists of different lengths. Now use pure functions to
• calculate the lengths of the sublists,
• throw out the first element of each sublist,
• rotate all sublists to the left by two elements,
• determine the sum of the elements in each sublist.

- **Dot and Cross Products**

Define your own functions to calculate the scalar and the cross product of two vectors.

- **A Riddle**

What does the following function do when it is applied to two lists of length 3:

$$\texttt{riddle[u\_, v\_] := RotateLeft[u RotateLeft[v] - RotateLeft[u] v]}$$

- **Sums**

What does the *Mathematica* function Sum give for $\sum_{i=1}^{n} \frac{1}{i^2 + i}$ and $\sum_{i=1}^{\infty} \frac{1}{i^3 + i^2}$?

- **Variations Using Apply**

Find at least three different ways of calculating the sum of the squares of all integers between 1 and 10000. Use the function Timing, to compare processing times.

- **Plots of Lists; Sums and Series**

Before calculating sums, look at the observations on efficiency in the in-depth section.

Plot the list of points $\frac{1}{i}$, $i = 1, \ldots, 500$.

Visualize the behavior of the sums $\sum_{i=1}^{n} \frac{1}{i}$, for $n = 1, \ldots, 500$, graphically.

The (infinite) harmonic series seems to diverge. What is *Mathematica*'s comment to this?

Compare with the behavior of $\sum_{i=1}^{n} \frac{1}{i^2}$.

Calculate the exact result for the series $\sum_{i=1}^{\infty} \frac{1}{i^2}$.

■ 3.3 Linear Algebra

Mathematica can also be used for linear algebra.

Several useful matrices are already pre-defined. IdentityMatrix gives us the identity matrix of the selected dimension.

```
In[92]:= IdentityMatrix[3]
```

```
Out[92]= {{1, 0, 0}, {0, 1, 0}, {0, 0, 1}}
```

DiagonalMatrix simplifies the definitions of diagonal matrices.

```
In[93]:= diag = DiagonalMatrix[{a, b, c}]
```

```
Out[93]= {{a, 0, 0}, {0, b, 0}, {0, 0, c}}
```

As we have already seen, `MatrixForm` produces a pretty print of matrices.

In[94]:= **MatrixForm[diag]**

Out[94]//MatrixForm=
$$\begin{pmatrix} a & 0 & 0 \\ 0 & b & 0 \\ 0 & 0 & c \end{pmatrix}$$

The inverse is calculated using `Inverse`.

In[95]:= **MatrixForm[Inverse[diag]]**

Out[95]//MatrixForm=
$$\begin{pmatrix} \frac{1}{a} & 0 & 0 \\ 0 & \frac{1}{b} & 0 \\ 0 & 0 & \frac{1}{c} \end{pmatrix}$$

The `.` operator calculates the matrix product. It is a short form of the `Dot` function.

In[96]:= **MatrixForm[% . diag]**

Out[96]//MatrixForm=
$$\begin{pmatrix} 1 & 0 & 0 \\ 0 & 1 & 0 \\ 0 & 0 & 1 \end{pmatrix}$$

`Dot` automatically sums over the last index of the first factor and the first index of the second factor. Therefore a list as first factor is a line vector and a list as second factor is a column vector.

In[97]:= **{a, b}.{{1, 2}, {1, 2}}**

Out[97]= $\{a + b,\ 2a + 2b\}$

In[98]:= **{{1, 2}, {1, 2}}.{a, b}**

Out[98]= $\{a + 2b,\ a + 2b\}$

The awkward differentiation between

In[99]:= **{{a, b}}.{{1, 2}, {1, 2}}**

Out[99]= $\{\{a + b,\ 2a + 2b\}\}$

In[100]:= **{{1, 2}, {1, 2}}.{{a}, {b}}**

Out[100]= $\{\{a + 2b\},\ \{a + 2b\}\}$

is usually unnecessary.

Now we work with a slightly more complicated 3×3 matrix.

In[101]:= **mat1 = {{a, c, 1}, {a, b, c}, {1, -b, 1}}**

Out[101]= {{a, c, 1}, {a, b, c}, {1, -b, 1}}

In[102]:= **MatrixForm[mat1]**

Out[102]//MatrixForm=
$$\begin{pmatrix} a & c & 1 \\ a & b & c \\ 1 & -b & 1 \end{pmatrix}$$

In[103]:= **MatrixForm[Inverse[mat1]]**

Out[103]//MatrixForm=
$$\begin{pmatrix} \frac{b+bc}{-b-ac+abc+c^2} & \frac{-b-c}{-b-ac+abc+c^2} & \frac{-b+c^2}{-b-ac+abc+c^2} \\ \frac{-a+c}{-b-ac+abc+c^2} & \frac{-1+a}{-b-ac+abc+c^2} & \frac{a-ac}{-b-ac+abc+c^2} \\ \frac{-b-ab}{-b-ac+abc+c^2} & \frac{ab+c}{-b-ac+abc+c^2} & \frac{ab-ac}{-b-ac+abc+c^2} \end{pmatrix}$$

In[104]:= **MatrixForm[Simplify[%.mat1]]**

Out[104]//MatrixForm=
$$\begin{pmatrix} 1 & 0 & 0 \\ 0 & 1 & 0 \\ 0 & 0 & 1 \end{pmatrix}$$

The transposed matrix is calculated using Transpose

In[105]:= **MatrixForm[Transpose[mat1]]**

Out[105]//MatrixForm=
$$\begin{pmatrix} a & a & 1 \\ c & b & -b \\ 1 & c & 1 \end{pmatrix}$$

the determinant using Det.

In[106]:= **Det[mat1]**

Out[106]= $-b - ac + abc + c^2$

■ In Depth

• An Application of Transpose

The function Transpose can be useful for problems which have little to do with linear algebra.

For example, let us look at a list of data which might come from an experiment and have been read with ReadList.

```
In[107]:=  expData = Table[N[Exp[-t] Cos[t]], {t, 0, 3, .3}]
```

```
Out[107]=  {1., 0.707731, 0.452954, 0.252728, 0.10914, 0.0157836,
            -0.0375563, -0.0618217, -0.0668948, -0.0607586, -0.0492888}
```

The corresponding values of the variable t are also given as list:

```
In[108]:=  tValues = Range[0, 3, .3]
```

```
Out[108]=  {0, 0.3, 0.6, 0.9, 1.2, 1.5, 1.8, 2.1, 2.4, 2.7, 3.}
```

To create a list of matching pairs with ListPlot we could of course iterate as follows:

```
In[109]:=  listPlotdata =
             Table[{tValues[[i]], expData[[i]]}, {i, Length[tValues]}]
```

```
Out[109]=  {{0, 1.}, {0.3, 0.707731}, {0.6, 0.452954},
            {0.9, 0.252728}, {1.2, 0.10914}, {1.5, 0.0157836},
            {1.8, -0.0375563}, {2.1, -0.0618217},
            {2.4, -0.0668948}, {2.7, -0.0607586}, {3., -0.0492888}}
```

```
In[110]:=  ListPlot[listPlotdata];
```

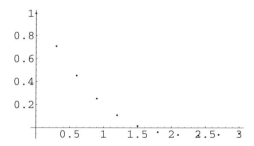

However this is much more elegant:

In[111]:= **ListPlot[Transpose[{tValues, expData}]];**

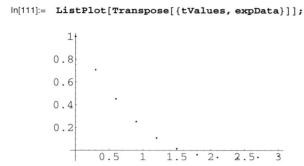

■ Exercises

● A Set of Simultaneous Equations

Let us look at the set of equations $\{2x + 5y + z = 1, 3x - y - z = 2, x + 5y + 3z = 1\}$. Solve it in two ways:

• with the function `Solve`,

• by determining the coefficient matrix and calculating the result using its inverse.

● Reading Programs

The steps below automate the above calculation. Try to understand them.

It is not necessary that the terms with the variables are on the left-hand side of the equations and the constant vector on the right. Therefore we first put everything on the left-hand sides and use this as a list.

In[112]:= **equations = {2 x + 5 y + z == 1, 3 x - y - z == 2, x + 5 y + 3 z == 1};**

In[113]:= **lhs = (#⟦1⟧ - #⟦2⟧ &) /@ equations**

Out[113]= $\{-1 + 2\,x + 5\,y + z,\ -2 + 3\,x - y - z,\ -1 + x + 5\,y + 3\,z\}$

Here we can determine the coefficient matrix, using the `Coefficient` function, as

In[114]:= **Coefficient[#, {x, y, z}] & /@ lhs**

Out[114]= $\{\{2,\ 5,\ 1\},\ \{3,\ -1,\ -1\},\ \{1,\ 5,\ 3\}\}$

We obtain the vector by setting the variables to zero. `Thread` does this the fastest. (What does `Thread[{x,y,z}→{0,0,0}]` do? What does `Thread[{x,y,z}→0]` do?)

In[115]:= **lhs /. Thread[{x, y, z} → 0]**

Out[115]= $\{-1,\ -2,\ -1\}$

Only the sign of the vector needs to be changed, because we have written everything on the left-hand side of the equation. This allows us to define the following two functions:

```
In[116]:=  coefficientMatrix[equations_, vars_] :=
           (Coefficient[#, vars] &) /@ (#[[1]] - #[[2]] &) /@ equations
```

```
In[117]:=  vector[equations_, vars_] :=
           (#[[2]] - #[[1]] &) /@ equations /. Thread[vars → 0]
```

```
In[118]:=  coefficientMatrix[{2 x + 5 y + z == w - 1,
             3 x - y - z - w + 2 == 0, 1 == x + 5 y + 3 z, w + x == 2}, {x, y, z, w}]
```

```
Out[118]=  {{2, 5, 1, -1}, {3, -1, -1, -1}, {-1, -5, -3, 0}, {1, 0, 0, 1}}
```

```
In[119]:=  vector[{2 x + 5 y + z == w - 1, 3 x - y - z - w + 2 == 0,
             1 == x + 5 y + 3 z, w + x == 2}, {x, y, z, w}]
```

```
Out[119]=  {-1, -2, -1, 2}
```

■ 3.4 Graphics Programming

■ 3.4.1 Graphics Objects

Mathematica recognizes various two-dimensional *graphics primitives*: Point, Line, Rectangle, Polygon, Circle, Disk, Raster, and Text. We can create a list containing such graphics primitives, use it in a Graphics object, and draw the illustration with Show.

```
In[120]:=  Show[Graphics[
             {Line[{{0, 0}, {1, 1}}],
              Circle[{0, 0}, Sqrt[2]],
              Text["Radius", {.8, .4}]}]];
```

All Graphics options, which we have already met as additional options of Plot, can be used to change the default values.

In[121]:= **Show[%, Axes -> True, AspectRatio -> 1, ImageSize -> 180];**

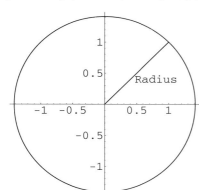

Graphics directives define properties of graphics primitives using, for instance, color (Hue, etc.), different point sizes (PointSize and AbsolutePointSize), thicknesses (Thickness and AbsoluteThickness), and dashes (Dashing and Absolute-Dashing).

For graphics directives which begin with Absolute, the value is given in pixels, those without Absolute use the percentage of the width of the graphic.

Graphics directives are valid for the successive elements of the list in which they appear and for its sublists.

Here AbsoluteThickness has an effect on all the rest:

In[122]:= **Show[Graphics[**
            {AbsoluteThickness[3],
             Line[{{0, 0}, {1, 1}}],
             Circle[{0, 0}, Sqrt[2]],
             Text["Radius", {.8, .4}]}],
            Axes -> True, AspectRatio -> 1, ImageSize -> 180];

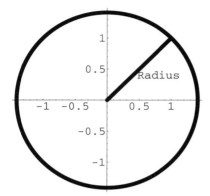

This way the "radius" is not drawn thickly:

In[123]:= **Show[Graphics[**
            {Line[{{0, 0}, {1, 1}}],
             AbsoluteThickness[3],
             Circle[{0, 0}, Sqrt[2]],
             Text["Radius", {.8, .4}]}],
            Axes -> True, AspectRatio -> 1, ImageSize -> 180];

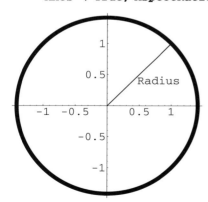

Now we enter the radius and the circle in a sublist, use AbsoluteThickness on it, and then draw a small thin circle in the middle.

```
In[124]:=  Show[Graphics[
                {{AbsoluteThickness[3],
             Line[{{0, 0}, {1, 1}}], Circle[{0, 0}, Sqrt[2]]},
             Circle[{0, 0}, .05], Text["Radius", {.8, .4}]}],
           Axes -> True, AspectRatio -> 1, ImageSize -> 180];
```

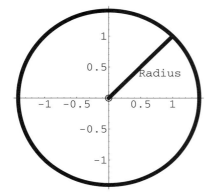

■ 3.4.2 Graphics3D Objects

Everything works the same way with three-dimensional graphics objects. Cuboid, Line, Point, Polygon, and Text are available as graphics primitives. Lists of such primitives, perhaps together with graphics directives, are written into a Graphics3D object and drawn using Show.

As an example, we draw 100 randomly placed cubes. To create the list of corner points, we use the (pseudo) random generator Random. In its simplest call version, without an argument, the results lie in the interval [0,1]. Therefore

```
In[125]:=  Table[{Random[], Random[], Random[]}, {100}];
```

returns a list of 100 "random" number triples. The Cuboid primitives are unit cubes with given corner points (as long as only one argument is given). We scale the coordinates by factor 20 so that the cubes will not all overlap. We can map Cuboid onto the list of coordinates using Map or /@ and put the results into a Graphics3D object.

In[126]:= **Show[Graphics3D[Cuboid /@**
          (20 Table[{Random[], Random[], Random[]}, {100}])],
          ImageSize -> 180];

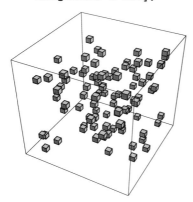

For a random colorization of the cubes (with `SurfaceColor`) we use the function `Transpose`, which was discussed in the above in-depth section.

In[127]:= **Show[Graphics3D[**
          Transpose[{Table[SurfaceColor[Hue[Random[]]], {100}],
          Cuboid /@ (20 Table[{Random[], Random[], Random[]},
          {100}])}]], ImageSize -> 180];

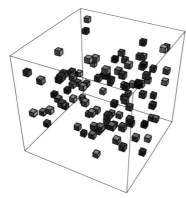

■ 3.4.3 Objects from Standard Packages

Several standard packages collected in the `Graphics` directory contain tools to create graphics objects. We load the definitions using

In[128]:= **<< Graphics`**

▪ Arrows

The Graphics`Arrow` package contains graphics objects for arrows. We can use them like this:

In[129]:= **Show[Graphics[{Arrow[{0, 0}, {1, 1}], Arrow[{0, 1}, {1, 0}]}],**
          ImageSize → 160];

The documentation in the *Help Browser* explains the various options for changing arrow-heads.

In[130]:= **Show[Graphics[{Arrow[{0, 0}, {1, 1}, HeadLength → 0.1],**
          Arrow[{0, 1}, {1, 0}, HeadLength → 0.1,
          HeadCenter → 0]}], ImageSize → 160];

▪ Polyhedra

We find the definitions of polyhedra in the Graphics`Polyhedra` package. As an example, let us draw an icosahedron.

In[131]:= **Show[Polyhedron[Icosahedron],**
        Boxed -> False, ImageSize -> 180];

▪ Three-Dimensional Objects

The Graphics`Shapes` package contains definitions for cylinders, cones, tori, sphere surfaces, etc., for use in Graphics3D objects.

In[132]:= **Show[Graphics3D[MoebiusStrip[3, 1, 50]],**
        Boxed -> False, ImageSize -> 180];

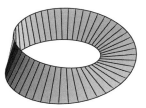

There are also functions for rotating objects and for drawing grid models.

In[133]:= **Show[Graphics3D[Sphere[1, 20, 20]],**
        Boxed -> False, ImageSize -> 180];

In[134]:= **Show[WireFrame[Graphics3D[Sphere[1, 20, 20]]],**
        Boxed -> False, ImageSize -> 180];

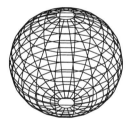

■ In Depth

• Splines

Roughly speaking, *spline* functions approximate polygons with smooth polynomial curves. There are various ways of doing this, which you can select for each problem accordingly. Many spline versions are pre-defined in the Graphics`Spline` package.

Let us look at a couple of points and the corresponding polygon.

In[135]:= **points = {{0, 0}, {0, 1}, {1, 1}, {2, 2}};**

In[136]:= **Show[Graphics[{Hue[0], Line[points]}], ImageSize -> 160];**

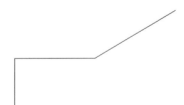

A cubic interpolation is drawn using the following command:

In[137]:= **Show[Graphics[{Hue[0], Line[points],**
          GrayLevel[0], Spline[points, Cubic]}], ImageSize -> 160];

■ Exercises

● The Pythagorean Theorem

Draw a right-angled triangle and the three squares of the Pythagorean theorem on its sides.
Hint: use AspectRatio, to scale the drawing properly.

Color the squares differently.

● The Thales Circle

Draw a right-angled triangle and the corresponding Thales circle.

Indicate the center of the circle with a small dot.

In addition, draw a radius with an arrow.

Label the sides of the triangle, the radius, and the circle.

- **Tori**

The `Torus` function (from `Graphics`Shapes``) produces a long list of polygons. Throw out the first 40 elements of the list and look at the resulting object.

Rotate the picture, so that you can see into the hole.

Draw a grid model of it.

Draw a torus with a 24×12 grid.

Drop the appropriate polygons from the list so that you get the torus with holes shown on the title page of Part 1.

If you double the number of polygons in every direction and color the surfaces, you get the title picture of the book.

- **Combining Objects**

Using `Graphics`Shapes`` draw a sphere and a long enough cylinder with half the radius of the sphere and its axis through the middle of the sphere. The result could look like this:

Now use the `TranslateShape` function (from the package) to move the cylinder by one radius length in direction x.

Draw a grid model of this object.

- **Boxes**

The title illustration to Part 4 consists of open "boxes" with five polygons making up the sides. It is quite easy to place such boxes on a virtual sphere by using a parametrization with spherical coordinates. Try it.

■ 3.5 Application: Animating a Mechanism

Now we will use our knowledge to animate a simple plane mechanism. This consists of two rods (1 and 4 units long) connected by a cylindrical joint. The short rod is supported by a cylindrical joint and rotates with a constant angular velocity. The long one slides on a horizontal plane. The support is 2 units above the horizontal plane.

(If the graphics look jaggy on the screen, redraw them smoothly with **Cell > Rerender Graphics**.)

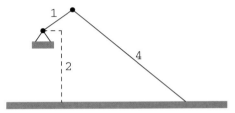

We want to vary the angle of rotation φ.

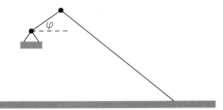

If we place the origin of our coordinate system into the support, we can determine the coordinates of the rotating joint:

In[138]:= **joint[φ_] = {Cos[φ], Sin[φ]};**

The sliding end of the long rod has the y coordinate -2. The x coordinate is comprised of that of the joint and the horizontal side of the large right-angled triangle, which we calculate using the Pythagorean theorem.

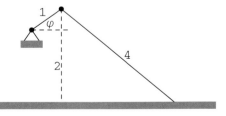

In[139]:= **endPoint[φ_] = {Cos[φ] + $\sqrt{16 - (2 + \text{Sin}[φ])^2}$, -2};**

This allows us to define a function which yields, at a given angle $φ$, the graphics primitives for rendering both rods and a circle for the joint.

In[140]:= **rods[φ_] =**
    {Disk[joint[φ], .07], Line[{{0, 0}, joint[φ], endPoint[φ]}]};

In[141]:= **Show[Graphics[rods[0]],**
    AspectRatio -> Automatic, ImageSize -> 160];

Now we define a list of objects to illustrate the supports. They are fixed, therefore we do not need a function of $φ$.

In[142]:= **supports =**
    {Line[{{-.2, -.3}, {0, 0}, {.2, -.3}}], Disk[{0, 0}, .07],
    GrayLevel[.5], Rectangle[{-.3, -.3}, {.3, -.5}],
    Rectangle[{-1, -2}, {5, -2.2}]};

In[143]:= **Show[Graphics[supports], AspectRatio -> Automatic];**

Now we are almost finished. We use `Table` to create a list of all the graphics and make sure that the same range is drawn for each angle.

In[144]:= **Table[Show[Graphics[{supports, rods[φ]}],**
       PlotRange → {{-1.1, 5.1}, {-2.5, 1.5}},
       AspectRatio → Automatic], {φ, 0, 2 π - $\frac{π}{10}$, $\frac{π}{10}$}];

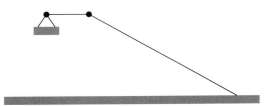

This cell group can be closed in the notebook and animated with **Cell > Animate Selected Graphics**.

The following command shows all the pictures of the "movie" overlapping:

In[145]:= **Show[%];**

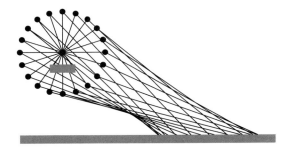

■ Exercises

• Sketches

Create the three sketches of the mechanism from the above section.

- **Parabola**

Animate the flight of a particle in a uniform gravitational field, neglecting the friction.

Hint: If the flight starts at the origin with an initial velocity v_0, the initial angle is α, and the acceleration of gravity g, then the x and y coordinates at time t are given by $x = v_0 \, t \, (\cos \alpha)$ and $y = v_0 \, t \, (\sin \alpha) - \frac{g \, t^2}{2}$.

The notebook contains a simple solution:

Superimpose a plot of the corresponding parabola.

Here is a possible solution for this as well:

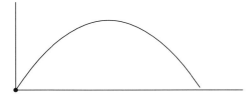

- **Cycloids**

A wheel rolls on a horizontal surface. At half radius a point is marked on the wheel. As the wheel moves this indicates a (shortened) cycloid. Visualize the wheel and the curve.

A static picture could look like this:

The notebook contains a suggestion for the animation:

This gives you a starting point for animations of more generalized cycloids.

Part 4: Introduction to Programming

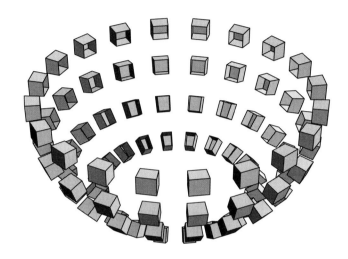

Part 4 discusses in greater detail how *Mathematica* works, so that we can tailor our calculations and develop simple programs. To do this, we need to understand the internal representation of expressions and know how patterns are used in definitions and transformation rules.

We will see that *Mathematica* contains all the tools for the well-known procedural programming styles in languages like Pascal, Modula-2, C, or Fortran, but that many problems can be solved much simpler using functional or rule-based programming.

The conclusion contains links to literature and to programs available on the World Wide Web.

■ 4.1 Expressions

Mathematica works internally with a uniform representation of all objects as *expressions*. Even entire notebooks are expressions and can be programmed accordingly (see Section 2.10.3 of the *Mathematica Book*). Numbers, names, and strings are atomic expressions. Nonatomic expressions have the form $f[a_1, a_2, \ldots]$. The name f denotes the *head*, and zero or more *arguments* a_1, a_2, ... are placed in square brackets. The arguments are themselves expressions.

The function FullForm shows the representation as an expressions. With

    In[1]:= **FullForm[f[x]]**

    Out[1]//FullForm=
          f[x]

nothing sensational happens, because f[x] was already written as an expression. More interesting is:

    In[2]:= **FullForm[(a + b)^n]**

    Out[2]//FullForm=
          Power[Plus[a, b], n]

Here we see that a+b is written internally as Plus[a,b] and the power with Power. An alternative view is produced by TreeForm, which displays the hierarchical structure as a tree.

    In[3]:= **TreeForm[(a + b)^n]**

    Out[3]//TreeForm=
          Power[| , n]
                  Plus[a, b]

We see that the first argument of Power is itself a nonatomic expression (that is Plus[a,b]); the vertical line | indicates the next level in the hierarchy. The second argument of Power is atomic.

Somewhat more complicated is:

```
In[4]:=  TreeForm[{a, (a^2 - b)^n}]
```

Out[4]//TreeForm=

$$
\text{List}\big[a, \;|\hspace{8cm}\big]
$$
$$
\text{Power}\big[\;|\hspace{5cm}, n\big]
$$
$$
\text{Plus}\big[\;|\hspace{2cm}, \;|\hspace{2cm}\big]
$$
$$
\text{Power}[a, 2]\quad \text{Times}[-1, b]
$$

This internal representation of expressions is important with many problems (see in-depth). *Mathematica* uses it in all calculations, especially in pattern recognition.

Many functions which we know from list manipulation also work for expressions. Have a look at:

```
In[5]:=  expr = 1 + x + x^2
```

$$
\text{Out[5]}= \; 1 + x + x^2
$$

```
In[6]:=  FullForm[expr]
```

Out[6]//FullForm=

```
        Plus[1, x, Power[x, 2]]
```

The first part (the first argument of the outer expression) is

```
In[7]:=  expr[[1]]
```

Out[7]= 1

the second part is

```
In[8]:=  expr[[2]]
```

Out[8]= x

We can easily add a further element using `Append`:

```
In[9]:=  Append[expr, x^3]
```

$$
\text{Out[9]}= \; 1 + x + x^2 + x^3
$$

■ In Depth

• Patterns for Rational and Complex Numbers

For rational numbers *Mathematica* uses a representation with `Rational`:

In[10]:= **FullForm[3 / 4]**

Out[10]//FullForm=
```
        Rational[3, 4]
```

To determine the numerator and the denominator of a rational number and combine the two as a list, we write:

In[11]:= **numAndDen[Rational[a_, b_]] = {a, b}**

Out[11]= {a, b}

In[12]:= **numAndDen[3 / 4]**

Out[12]= {3, 4}

The following version does not work:

In[13]:= **wrong[a_ / b_] = {a, b}**

Out[13]= {a, b}

In[14]:= **wrong[3 / 4]**

Out[14]= wrong$\left[\dfrac{3}{4}\right]$

because the expression wrong[Rational[3,4]] is not matched by the pattern in the definition:

In[15]:= **FullForm[wrong[a_ / b_]]**

Out[15]//FullForm=
```
        wrong[Times[Pattern[a, Blank[]], Power[Pattern[b, Blank[]], -1]]]
```

We see this more clearly without the blanks:

In[16]:= **FullForm[wrong[a / b]]**

Out[16]//FullForm=
```
        wrong[Times[a, Power[b, -1]]]
```

In the same way, complex numbers are written internally using Complex:

In[17]:= **FullForm[2 + 3 I]**

Out[17]//FullForm=
```
        Complex[2, 3]
```

■ Exercises

• The Structure of Expressions

Study the internal representations of the following expressions:

```
(a + b) ^ 2
```

$$a^2 + 2\,a\,b + b^2$$

```
x'[t]
```

```
D[s[x, y], x, y]
```

- **Real and Imaginary Parts**

Study the above in-depth section. Using pattern recognition, define a function which returns the real and imaginary parts of a complex number as a list. (Do not use `Re` or `Im`.)

■ 4.2 Patterns

We have already seen in the first part that the left-hand side of transformation rules and definitions must be interpreted as *patterns*. The patterns generally contain *blanks* (_) which can be filled with any expression. So a pattern matches an expression if the expression has exactly the same structure as the pattern (in the internal representation), but any sub-expressions can appear instead of blanks in the pattern.

There are various useful tools to restrict patterns or construct more complicated patterns.

■ 4.2.1 Simple Patterns

Let us look at the following expression:

In[18]:= **formula = 1 + x + x ^ 2 + y ^ 3 + z ^ 2 + x ^ 2 Sin[z]**

Out[18]= $1 + x + x^2 + y^3 + z^2 + x^2\,\text{Sin}[z]$

We can substitute values with the help of transformation rules.

In[19]:= **formula /. x -> 3**

Out[19]= $13 + y^3 + z^2 + 9\,\text{Sin}[z]$

The left-hand side of the transformation rule, i.e. x, is in this case a very special pattern. It only matches the expression x. If we replace x with a `Blank` (_), this pattern matches the entire formula and everything is replaced by the right-hand side of the transformation rule.

In[20]:= **formula /. _ -> 3**

Out[20]= 3

It gets more interesting if we use a _^2 pattern to set all squares to zero.

In[21]:= **formula /. _^2 -> 0**

Out[21]= $1 + x + y^3$

Or we can make all powers disappear:

In[22]:= **formula /. _^_ -> 0**

Out[22]= $1 + x$

Or we can rewrite a sum of two squares in a new form:

In[23]:= **formula /. a_^2 + b_^2 -> sumOfSquares[a, b]**

Out[23]= $1 + x + y^3 + x^2 \operatorname{Sin}[z] + \operatorname{sumOfSquares}[x, z]$

The use of patterns in definitions is completely analogous. For example, we can extract the coefficients of a linear polynomial as a list:

In[24]:= **coeffs[a_ + b_ x_, x_] = {a, b};**

In[25]:= **coeffs[1 + 2 y, y]**

Out[25]= {1, 2}

If an expression is not matched by the pattern, it will not be evaluated.

In[26]:= **coeffs[1 + 2 x + 4 y^2, y]**

Out[26]= $\operatorname{coeffs}[1 + 2 x + 4 y^2, y]$

In order to really be useful, the above definition must be improved. In the following cases it does not work like we want it to:

In[27]:= **coeffs[2 y, y]**

Out[27]= $\operatorname{coeffs}[2 y, y]$

(The expression is not matched by the pattern because there is no constant summand.)

In[28]:= **coeffs[1 + y, y]**

Out[28]= coeffs[1 + y, y]

(The expression is not matched by the pattern because there is no factor in the linear term.)

In[29]:= **coeffs[1 + 2 y + y^2, y]**

Out[29]= {1 + y^2, 2}

(1+y^2 is matched by the a_ of the pattern.)

Luckily there are simple tools to handle such cases. They will be discussed in the next two sections.

■ 4.2.2 Constraints

There are three ways of defining patterns that only match under constraints:
• fixing the "type" (head) of the expression,
• constraints with /; operators,
• constraints with test functions.

■ Constraints on Heads

We have seen that each expression has a *head*, which can be interpreted as the *type* of the expression.

The function Head shows that even atomic expressions have heads which are normally hidden:

In[30]:= **Head /@ {a, "x", 1, 1.1}**

Out[30]= {Symbol, String, Integer, Real}

The head of a list is List.

In[31]:= **Head[{a, b}]**

Out[31]= List

A blank followed by the name of the desired head only matches expressions with that head.

As an example for this technique, let us look at a function which returns the first element of a list. (For this we need a delayed definition because the right-hand side can only be evaluated once the list has been inserted.)

```
In[32]:= firstElement[l_] := l[[1]]
```

```
In[33]:= firstElement[{a, b, c}]
```

Out[33]= a

This definition produces an error message if we apply it to an atomic expression.

```
In[34]:= firstElement[1]
```

```
        Part::partd :
           Part specification 1[[1]] is longer than depth of object.
```

Out[34]= 1[[1]]

The following improved version restricts the pattern to arguments of type List. It will avoid the error messages:

```
In[35]:= Clear[firstElement]
```

```
In[36]:= firstElement[l_List] := l[[1]]
```

```
In[37]:= firstElement[{a, b, c}]
```

Out[37]= a

```
In[38]:= firstElement[1]
```

Out[38]= firstElement[1]

■ Constraints with /;

With the operator /; constraints can be applied to the pattern itself or to the entire definition. On the right-hand side of the operator there must be a test which yields the result True for those expressions which should be matched by the pattern.

This gives us a function which only evaluates positive arguments:

```
In[39]:= numRoot[x_] := √N[x] /; x ≥ 0
```

```
In[40]:= numRoot[2]
```

Out[40]= 1.41421

In[41]:= **numRoot[-1]**

Out[41]= numRoot[-1]

The alternative version, in which the constraint is placed directly into the pattern for the arguments, works just as well.

In[42]:= **numRoot2[x_ /; x ≥ 0] := $\sqrt{N[x]}$**

In[43]:= **numRoot2 /@ {2, -1}**

Out[43]= {1.41421, numRoot2[-1]}

The following example can only be defined using a constraint on the definition:

In[44]:= **rootOfSum[x_, y_] := $\sqrt{N[x+y]}$ /; x + y ≥ 0**

In[45]:= **rootOfSum[5, -2]**

Out[45]= 1.73205

In[46]:= **rootOfSum[-5, 2]**

Out[46]= rootOfSum[-5, 2]

We can also create piecewise functions with this kind of constraint.

In[47]:= **piecewise[x_] := x² /; x > 0**

In[48]:= **piecewise[x_] := -x /; x ≤ 0**

In[49]:= **Plot[piecewise[x], {x, -1, 1}];**

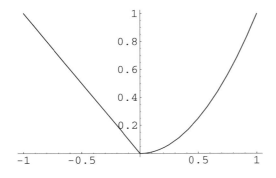

Derivatives and integrals of functions which were defined in this way cannot be evaluated. For such tasks it is preferable to use the function `UnitStep`, which in Version 3.0.x is in the `Calculus`DiracDelta`` package and will be built into the kernel in future versions.

■ Test Functions

Mathematica contains a great number of test functions which are useful for constraints on patterns. Their name always ends with Q. They result in `True` if the test has been successful or in `False` in all other cases: if the test has been unsuccessful or if the result is undetermined. The following command lists all test functions:

```
In[50]:=   ? *Q
```

```
ArgumentCountQ          MatrixQ
AtomQ                   MemberQ
DigitQ                  NameQ
EllipticNomeQ           NumberQ
EvenQ                   NumericQ
ExactNumberQ            OddQ
FreeQ                   OptionQ
HypergeometricPFQ       OrderedQ
InexactNumberQ          PartitionsQ
IntegerQ                PolynomialQ
IntervalMemberQ         PrimeQ
InverseEllipticNomeQ    SameQ
LegendreQ               StringMatchQ
LetterQ                 StringQ
LinkConnectedQ          SyntaxQ
LinkReadyQ              TrueQ
ListQ                   UnsameQ
LowerCaseQ              UpperCaseQ
MachineNumberQ          ValueQ
MatchLocalNameQ         VectorQ
MatchQ
```

The functions `Positive`, `Negative`, and `NonNegative` are also useful. However, they may remain unevaluated:

```
In[51]:=   Positive /@ {-1, 0, 1, a}
```

```
Out[51]=   {False, False, True, Positive[a]}
```

Using `TrueQ` we can create test functions which in such a case give a `False`:

```
In[52]:=   myPositiveQ[x_] := TrueQ[Positive[x]]
```

```
In[53]:=   myPositiveQ /@ {-1, 0, 1, a}
```

```
Out[53]=   {False, False, True, False}
```

With the function `FreeQ`, which tests if an expression contains a symbol, we can rule out in the above definition of `coeffs` that the "constant summand" contains a higher power of x. We must first clear the old definition, which would otherwise remain active, and use a delayed definition, so that the test is evaluated correctly.

```
In[54]:= Clear[coeffs]
```

```
In[55]:= coeffs[a_ + b_ x_, x_] := {a, b} /; FreeQ[a, x]
```

```
In[56]:= coeffs[1 + 2 y + y², y]
```

```
Out[56]= coeffs[1 + 2 y + y², y]
```

■ Constraints Using Test Functions

Patterns can also be constrained using (pure) test functions. For this we use the form *pattern?test*. The test function is then applied to the argument and the pattern matches only if the test function returns `True`. With `EvenQ` we restrict the arguments to be even:

```
In[57]:= half[n_ ? EvenQ] = n / 2;
```

```
In[58]:= half /@ {1, 2}
```

```
Out[58]= {half[1], 1}
```

We can define our own test functions, preferably as pure functions. The following definition tests whether the argument is a non-negative integer.

```
In[59]:= myFactorial1[n_ ? (# ≥ 0 && IntegerQ[#] &)] = n!;
```

```
In[60]:= myFactorial1 /@ {-1, 1 / 2, 10}
```

$$Out[60]= \left\{myFactorial1[-1], myFactorial1\left[\frac{1}{2}\right], 3628800\right\}$$

An alternative would be:

```
In[61]:= myFactorial2[n_ ? (NonNegative[#] && IntegerQ[#] &)] = n!;
```

```
In[62]:= myFactorial2 /@ {-1, 1 / 2, 10}
```

$$Out[62]= \left\{myFactorial2[-1], myFactorial2\left[\frac{1}{2}\right], 3628800\right\}$$

■ 4.2.3 More Complicated Patterns

There are several methods to create even more complicated patterns (see Section 2.3.6 *ff* of the *Mathematica Book*). We will only discuss the most important ones here.

■ Alternatives

With | we can combine different patterns.

```
In[63]:= x + x^2 + x^3 + y /. x | x^_ -> c
```

```
Out[63]= 3 c + y
```

■ Optional Arguments and Default Values

The following function adds its two arguments:

```
In[64]:= add[x_, y_] = x + y;
```

```
In[65]:= add[a, b]
```

```
Out[65]= a + b
```

In case of only one argument, the pattern will not match and therefore nothing happens.

```
In[66]:= add[a]
```

```
Out[66]= add[a]
```

But perhaps in such a case the argument itself should be returned. We obtain this by entering a default value after a colon. The default value is then used if the argument is missing.

```
In[67]:= Clear[add]
```

```
In[68]:= add[x_, y_: 0] = x + y;
```

```
In[69]:= add[a, b]
```

```
Out[69]= a + b
```

```
In[70]:= add[a]
```

```
Out[70]= a
```

For sums, products, and powers the default values 0, 1, and 1 are already built in; an optional argument is defined by placing a dot behind the blank. In this way we can create an improved variation of our `coeffs` function:

```
In[71]:= Clear[coeffs]
```

```
In[72]:= coeffs[a_. + b_. x_, x_] := {a, b} /; FreeQ[a, x]
```

A missing constant summand is now replaced by 0 and a missing coefficient in the linear term is replaced by 1.

```
In[73]:= coeffs[x, x]
```

```
Out[73]= {0, 1}
```

■ 4.2.4 A Simple Integrator

Because *Mathematica*'s evaluation process basically uses all definitions and rewrites the result until nothing more changes, we can easily create programs that can solve rather complicated tasks.

An an example, we program our own integration function for simple polynomial expressions (see Section 2.3.14 of the *Mathematica Book*). We call the function `toyInte-grate`. Just as with `Integrate` we suppose that the expression to be integrated is entered as the first argument and the variable as the second.

Linearity is treated by two definitions. The integral of a sum is the sum of the integrals:

```
In[74]:= toyIntegrate[y_ + z_, x_] :=
            toyIntegrate[y, x] + toyIntegrate[z, x]
```

Constants (which do not contain the function variables) can be drawn before the integral.

```
In[75]:= toyIntegrate[c_ y_, x_] := c toyIntegrate[y, x] /; FreeQ[c, x]
```

The integral of a constant is:

```
In[76]:= toyIntegrate[c_, x_] := c x /; FreeQ[c, x]
```

The integral of an integer power, except -1, can be processed using:

```
In[77]:= toyIntegrate[x_^n_., x_] := x^{n+1}/(n + 1) /; FreeQ[n, x] && n ≠ -1
```

These four definitions already do an amazing job.

In[78]:= **toyIntegrate$\left[\mathbf{a\,x^2 + b\,x + c + \dfrac{1}{x}},\ \mathbf{x} \right]$**

Out[78]= $c\,x + \dfrac{b\,x^2}{2} + \dfrac{a\,x^3}{3} + \text{toyIntegrate}\left[\dfrac{1}{x},\ x \right]$

The integral of $\frac{1}{x}$ cannot yet be determined. Nonetheless, the rest is automatically calculated as completely as possible.

With the additional definition

In[79]:= **toyIntegrate$\left[\dfrac{1}{\mathbf{a\_\,.\,x\_ + b\_\,.}},\ \mathbf{x\_} \right]$:=**

$\dfrac{\mathbf{Log[a\,x + b]}}{\mathbf{a}}$ **/; FreeQ[{a, b}, x]**

we come one step further:

In[80]:= **toyIntegrate$\left[\mathbf{a\,x^2 + b\,x + c + \dfrac{1}{x}},\ \mathbf{x} \right]$**

Out[80]= $c\,x + \dfrac{b\,x^2}{2} + \dfrac{a\,x^3}{3} + \text{Log}[x]$

The integrator in Version 1 of *Mathematica* was built up in this way, but since Version 2 it has been replaced by a much better algorithm.

■ In Depth

• Patterns of Derivatives

Derivatives have the following representation as *Mathematica* expressions:

In[81]:= **FullForm[x'[t]]**

Out[81]//FullForm=
    Derivative[1][x][t]

Because there is no x[t] in this expression (note the different bracketing), a transformation rule containing the pattern x[t] will not match derivatives. Such a rule is produced when solving a differential equation for x[t].

In[82]:= **DSolve[x'[t] == x[t], x[t], t]**

Out[82]= $\{\{x[t] \rightarrow E^t\,C[1]\}\}$

In[83]:= **x'[t] /. %[[1]]**

Out[83]= $x'[t]$

If we request the solution as a pure function, i.e. as a transformation rule for x itself, the pattern does match.

In[84]:= **DSolve[x'[t] == x[t], x, t]**

Out[84]= $\{\{x \to (E^{\#1} \, C[1] \, \&)\}\}$

In[85]:= **x'[t] /. %[[1]]**

Out[85]= $E^t \, C[1]$

In this way we can verify the solution.

In[86]:= **x'[t] == x[t] /. %%[[1]]**

Out[86]= True

• Several Arguments

Sometimes an unknown number of arguments should be processed, for instance in programming functions with options. Two blanks (__) stand for one or more arguments, three blanks (___) also include the case of no argument at all.

To illustrate this, let us look at a function that returns its arguments as a list. The list can also be empty, so we use three blanks:

In[87]:= **listOfArguments[x___] = {x}**

Out[87]= $\{x\}$

In[88]:= **listOfArguments[]**

Out[88]= $\{\}$

In[89]:= **listOfArguments[a, b, c]**

Out[89]= $\{a, b, c\}$

• Functions with Options

Now we want to develop the skeleton of a function with options. We call it skel. It shall have one argument and two options opt1 and opt2. To illustrate what is going on, the result shall be a list, consisting of the argument and the values of the two options.

The default values of the options shall be default1 and default2. If the user does not set an option, the default value will be used. It is a convention in *Mathematica* that the list of default values of options is passed to the built-in function Options in the following way:

In[90]:= **Options[skel] = {opt1 → default1, opt2 → default2};**

Now we can determine the default value of `opt1`:

```
In[91]:= opt1 /. Options[skel]
```

```
Out[91]= default1
```

Because several `/.` operators are evaluated from left to right, in the following expression the option `opt1` is first set to 3. Then the list of default values is applied to the result. However, this no longer has an effect on `opt1`, because it was already substituted by 3 before.

```
In[92]:= opt1 /. opt1 → 3 /. Options[skel]
```

```
Out[92]= 3
```

In this way the `skel` function can easily be defined:

```
In[93]:= skel[x_, opts___] := {x, opt1, opt2} /. {opts} /. Options[skel]
```

If no option values are given, the defaults are used:

```
In[94]:= skel[a]
```

```
Out[94]= {a, default1, default2}
```

But given options will be used:

```
In[95]:= skel[a, opt2 → myValue2]
```

```
Out[95]= {a, default1, myValue2}
```

Let us slightly improve the definition of `skel`. With the test function `OptionQ` we make sure that options (transformation rules) have actually been given. We also make sure it all works if the options are given as a list. The skeleton then finally looks like this:

```
In[96]:= Clear[skel]
```

```
In[97]:= skel[x_, opts___?OptionQ] :=
             {x, opt1, opt2} /. Flatten[{opts}] /. Options[skel]
```

```
In[98]:= skel[a]
```

```
Out[98]= {a, default1, default2}
```

```
In[99]:= skel[a, a]
```

```
Out[99]= skel[a, a]
```

```
In[100]:= skel[a, {opt1 -> myValue1, opt2 -> myValue2}]
```

```
Out[100]= {a, myValue1, myValue2}
```

Of course, this function does not do anything useful yet. In practice, the option values given by the user will probably be determined in a module, and the evaluation will be continued according to these values.

■ Exercises

• Gradients

Enhance the function `gradient` from the exercises to Section 3.2 so that it only evaluates if a list is entered as the second argument.

• Dot and Cross Products

Enhance the functions defined in the exercises to Section 3.2 for the calculation of dot and cross products, so that they are only evaluated for suitable inputs.

• Integrator

Enhance the integrator of Section 4.2.4 with a couple of additional definitions, for instance, to handle trigonometric functions.

The integration process can be illustrated using `Print` commands, so that whenever a definition is used a corresponding message is written. Put the right-hand side of the definition into parentheses, creating a compound expression:

```
toyIntegrate[y_ + z_, x_] := (Print["Sum rule for ", y, " and ", z];
   toyIntegrate[y, x] + toyIntegrate[z, x])
```

Use this method to enhance the definitions for `toyIntegrate`. Observe the evaluation of some examples.

The function can be improved even further by adding an option for switching messages on and off (see the above in-depth section). Branching with `If` is useful here.

■ 4.3 Evaluation

With this previous knowledge we can learn how *Mathematica* actually works. This will help us to single-mindedly develop our calculations and to understand why *Mathematica* sometimes returns unexpected results.

We begin with preliminaries about associated definitions and definitions with attributes and then observe the evaluation process of expressions

■ 4.3.1 Associated Definitions

Definitions are normally associated to the outermost head of the pattern. As necessary, they can also be associated with a head of an *argument*. This can be used to add properties of built-in functions.

We may wish to define the integral our own function myFunction:

In[101]:= **Integrate[myFunction[x_], x_] = integralOfMyFunction[x]**

Set::write : Tag Integrate in $\int$ myFunction[x_] dx_ is Protected.

Out[101]= integralOfMyFunction[x]

This does not work because the internal function Integrate is protected (by the attribute Protected).

(In principle, we could remove the protection of Integrate with Unprotect and then use the above definition. However this is very dangerous; an incorrect definition may render the integrator useless.)

But because the definition applies only to our function anyway, we can associate to it using the / : operator:

In[102]:= **myFunction /: Integrate[myFunction[x_], x_] = integralOfMyFunction[x]**

Out[102]= integralOfMyFunction[x]

In[103]:= **Integrate[myFunction[y], y]**

Out[103]= integralOfMyFunction[y]

A short form is ^= (and ^ : = for an associated, delayed definition):

In[104]:= **Integrate[yourFunction[x_], x_] ^= integralOfYourFunction[x]**

Out[104]= integralOfYourFunction[x]

■ 4.3.2 Attributes

Functions can also have *attributes* assigned to them in order to determine properties such as associativity, commutativity, or automatic mapping on lists. A complete list of all possible attributes can be found in the documentation to Attributes. This function shows us that Sin carries the attribute Listable.

In[105]:= **Attributes[Sin]**

Out[105]= {Listable, NumericFunction, Protected}

As a consequence `Sin` is mapped automatically on the elements of lists.

In[106]:= **Sin[{0, Pi/4, Pi/2}]**

Out[106]= $\left\{0, \dfrac{1}{\sqrt{2}}, 1\right\}$

Our own functions do not behave this way.

In[107]:= **isNotMapped[{0, Pi / 4, Pi / 2}]**

Out[107]= $\text{isNotMapped}\left[\left\{0, \dfrac{\pi}{4}, \dfrac{\pi}{2}\right\}\right]$

In[108]:= **Attributes[isNotMapped]**

Out[108]= $\{\}$

But we can assign the attribute `Listable` to define a function which is automatically mapped on lists:

In[109]:= **SetAttributes[isMapped, Listable]**

In[110]:= **isMapped[{0, Pi / 4, Pi / 2}]**

Out[110]= $\left\{\text{isMapped}[0], \text{isMapped}\left[\dfrac{\pi}{4}\right], \text{isMapped}\left[\dfrac{\pi}{2}\right]\right\}$

In[111]:= **Attributes[isMapped]**

Out[111]= $\{\text{Listable}\}$

■ 4.3.3 The Evaluation Process

Mathematica's evaluation process can be divided into three phases:
1. Reading the cell and transforming it into the internal representation as an expression.
2. Evaluating the expression.
3. Formatting the result for output.

Only the second step is of interest to us. Here all built-in and user-defined transformation rules and definitions are used to rewrite the expression, until nothing more changes. *Mathematica* does this in the following order:

2.1 Evaluation of the head.
2.2 Evaluation of each argument, in order.
2.3 Re-ordering using the attributes `Flat` (associative) and `Orderless` (commutative).

2.4 Mapping to lists (attribute `Listable`).

2.5 Application of user-defined definitions associated to the head of an argument.

2.6 Application of built-in definitions associated to the head of an argument.

2.7 Application of user-defined definitions associated to the head of the expression.

2.8 Application of built-in definitions associated to the head of the expression.

Through pattern recognition, the steps 2.5-2.8 are used to test whether the pattern of a rule or definition matches. If it does, the right-hand side of the definition is substituted and the evaluation restarts for the new expression.

After the evaluation of the head, step 2.2 of this process introduces a recursion, thereby finally evaluating the expression from the inside out. The recursion comes to an end in every branch of the tree (`TreeForm`) when atoms are reached: numbers, strings, and symbols without definitions evaluate to themselves; for a symbol with definitions, the right-hand side of the definition is evaluated in step 2.7.

This standard evaluation scheme can be changed (see Section 2.5.5 of the *Mathematica Book*), and several built-in functions must deviate from this in order to function properly. We will, however, not go into this further here.

`Trace` lists each step in the evaluation:

In[112]:= **`testFunction[x_, y_] := Simplify[x^2 - y^2]`**

In[113]:= **`someName = testFunction;`**

In[114]:= **`Trace[someName[Expand[(a + a + b)^2], a]]`**

Out[114]= $\{\{$someName, testFunction$\}$,
$\{\{\{$a + a + b, 2 a + b$\}$, $(2 a + b)^2\}$, Expand$[(2 a + b)^2]$,
$4 a^2 + 4 a b + b^2\}$, testFunction$[4 a^2 + 4 a b + b^2, a]$,
Simplify$\left[(4 a^2 + 4 a b + b^2)^2 - a^2\right]$,
$\{(4 a^2 + 4 a b + b^2)^2 - a^2, -a^2 + (4 a^2 + 4 a b + b^2)^2\}$,
Simplify$\left[-a^2 + (4 a^2 + 4 a b + b^2)^2\right]$, $-a^2 + (2 a + b)^4\}$

We see how first the head `someName` evaluates to `testFunction`. The evaluation of the first argument begins with the evaluation of the argument of `Expand`, then `Expand` itself is applied. Next, the definition for `testFunction` is used and its right-hand side is evaluated.

■ In Depth

• Delayed Transformation Rules

In addition to the (immediate) transformation rules with $->$ or $\to$ *Mathematica* also recognizes delayed ones. They are written as $:>$ or $:\to$. As with delayed definitions, the right-hand side is only evaluated after the pattern has been replaced.

In[115]:= $(\mathbf{a} + \mathbf{b})^2$ /. $\mathbf{x\_} \to \mathbf{Expand[x]}$

Out[115]= $(a + b)^2$

In[116]:= $(\mathbf{a} + \mathbf{b})^2$ /. $\mathbf{x\_} :\to \mathbf{Expand[x]}$

Out[116]= $a^2 + 2\, a\, b + b^2$

One can therefore understand an immediate definition to be a global immediate transformation rule, and a delayed definition to be a global delayed transformation rule.

• Repeated Application of Transformation Rules

A transformation rule is applied only once using / .. For repeated application until nothing more changes, we use the / / . operator. The difference is made clear in the following two expressions (see also Section 4.4.3).

In[117]:= **fac[5] /. {fac[0] -> 1, fac[n_] -> n fac[n - 1]}**

Out[117]= $5\,fac[4]$

In[118]:= **fac[5] //. {fac[0] -> 1, fac[n_] -> n fac[n - 1]}**

Out[118]= 120

• Hold

We cannot see an expression such as 1+1 in its full form at first, because according to the evaluation process in a

In[119]:= **FullForm[1 + 1]**

Out[119]//FullForm=
        2

the argument will be evaluated before FullForm is applied. For this the functions Hold and HoldForm are useful. They prevent the evaluation of their arguments:

In[120]:= **Hold[FullForm[1 + 1]]**

Out[120]= $Hold[Plus[1, 1]]$

In[121]:= **HoldForm[FullForm[1 + 1]]**

Out[121]= $Plus[1, 1]$

■ Exercises

● The Attribute `Orderless`

Study the documentation of the attribute `Orderless` and then interpret the evaluation of the following function:

```
SetAttributes[pr, Orderless]

pr[x___] := 1 /; Print[x]

pr[1, 2, 3]
```

● Application of Transformation Rules: Fibonacci Numbers

Fibonacci numbers can be calculated recursively: The zeroth is 0 and the first is 1, higher ones are the sum of the previous two. Use transformation rules to determine the tenth Fibonacci number.

(This works only for small numbers, because the number of calculations needed grows exponentially. See Section 4.4.3.)

■ 4.4 Programming Tools

We will now have a look at the most important tools for programming. This will show us that it is possible to program using different methodologies. Often the well-known procedural style, used in C, Fortran, or Pascal, is not the clearest nor the most efficient for *Mathematica*.

■ 4.4.1 Local Variables

If programs are to be given to users there lies the danger of collision between the names in the program and the names chosen by its user. There are two mechanisms for avoiding such name collisions in *Mathematica*. On the procedural or functional level, the `Module` mechanism is used to define local variables. The technique used on a global level, especially for the names of the functions themselves, will be discussed in Section 4.4.5.

The `Module` function has two arguments. The first is a list of local variables, the second a possibly compound expression (a series of single expressions divided by semicolons). Against intuition, commas divide more strongly than semicolons.

The following function calculates the rotation of a planar vector by an angle φ. It makes sense to avoid multiple calculations of trigonometric functions, because this is time-intensive. With the help of two local variables, we can write:

```
In[122]:=  rot2D[{x_, y_}, φ_] :=
              Module[{sinφ, cosφ},
                  sinφ = Sin[φ];
                  cosφ = Cos[φ];
                  {{cosφ, -sinφ}, {sinφ, cosφ}}.{x, y}
              ]
```

```
In[123]:=  rot2D[{1, 1}, Pi / 2]
```

```
Out[123]=  {-1, 1}
```

The local variables can be initialized as they are introduced by using immediate definitions. This gives us a more compact implementation.

```
In[124]:=  rot2D[{x_, y_}, φ_] := Module[{sinφ = Sin[φ], cosφ = Cos[φ]},
              {{cosφ, -sinφ}, {sinφ, cosφ}}.{x, y}]
```

If the result of a `Module` function is not calculated at the end of the module, it can be returned using `Return`.

The related functions `With` and `Block` will not be discussed here.

■ 4.4.2 Functional Programming

Mathematica is predestined for functional programming. This means the nesting of functions, as we have done a lot of times without thinking about it. Let us look at a class of problems which are well-suited as examples for functional programming.

Many nonlinear algorithms can be reduced to the search for fixed points of mappings. One starts at any initial point and applies the mapping. The same mapping is again applied to the result. And so on, until the difference between two successive results is within a chosen threshold.

A good example is *Newton's algorithm* for finding roots of functions. We consider a function of one variable and begin with an initial x value. At this point we draw the tangent to the graph and determine its intersection with the x axis. This gives us the first approximation of the root.

Let us choose the function

In[125]:= **f[x_] = Cos[x²] - Sin[x];**

In[126]:= **Plot[f[x], {x, 0, 2}];**

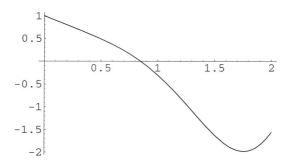

as an example. We need its derivative.

In[127]:= **df[x_] = ∂ₓ f[x]**

Out[127]= $-\text{Cos}[x] - 2 x \text{Sin}[x^2]$

The initial value shall be x_0=1.6. The tangent therefore looks like this:

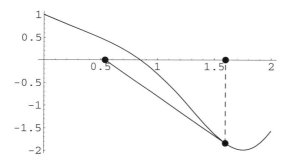

The intersection with the abscissa calculates as $x_0 - \frac{f(x_0)}{f'(x_0)}$. The corresponding pure function is:

In[128]:= **# - f[#] / df[#] &**

Out[128]= $\#1 - \frac{f[\#1]}{df[\#1]} \&$

Applied to the initial value, it yields:

In[129]:= **# - f[#] / df[#] &[1.6]**

Out[129]= 0.538438

With this value the same procedure is repeated to determine the second approximation of the root. With the exception of pathological cases (horizontal tangents, limit cycles) we very quickly arrive at a good approximation of the root. The *Mathematica* function `Nest` is very helpful for this process:

```
In[130]:= Nest[g, x, 5]
```

```
Out[130]= g[g[g[g[g[x]]]]]
```

`NestList` can be used to see the intermediate results as well:

```
In[131]:= NestList[g, x, 3]
```

```
Out[131]= {x, g[x], g[g[x]], g[g[g[x]]]}
```

In place of g we simply have to use our pure function–and the program is finished.

```
In[132]:= NestList[# - f[#] / df[#] &, 1.6, 5]
```

```
Out[132]= {1.6, 0.538438, 0.920372, 0.853035, 0.849379, 0.849369}
```

The nesting functions `FixedPointList` and `FixedPoint` work until two successive results agree (within a built-in threshold defined by `SameQ`):

```
In[133]:= FixedPointList[# - f[#] / df[#] &, 1.6]
```

```
Out[133]= {1.6, 0.538438, 0.920372, 0.853035,
           0.849379, 0.849369, 0.849369, 0.849369}
```

```
In[134]:= FixedPoint[# - f[#] / df[#] &, 1.6]
```

```
Out[134]= 0.849369
```

(In *Mathematica* 4, the functions `NestWhile` and `NestWhileList` give you extended control over the tests used to terminate the nesting.)

By using local variables, we could program Newton's algorithm as follows:

```
In[135]:= Clear[f, df]
```

```
In[136]:= myNewton[f_, {x_, x0_}] :=
              Module[{df = ∂_x f}, FixedPoint[#1 - f/df /. x → #1 &, N[x0]]]
```

```
In[137]:= myNewton[Cos[x^2] - Sin[x], {x, 1.6}]
```

```
Out[137]= 0.849369
```

The functions `FoldList`, `MapIndexed`, and `ComposeList` are also helpful in functional programs. They allow iterations over lists:

In[138]:= **FoldList[f, x, {a, b, c}]**

Out[138]= {x, f[x, a], f[f[x, a], b], f[f[f[x, a], b], c]}

In[139]:= **MapIndexed[f, {a, b, c}]**

Out[139]= {f[a, {1}], f[b, {2}], f[c, {3}]}

In[140]:= **ComposeList[{f1, f2, f3}, x]**

Out[140]= {x, f1[x], f2[f1[x]], f3[f2[f1[x]]]}

■ 4.4.3 Rule-Based and Recursive Programming

We saw a good example of rule-based programming in the integrator of Section 4.2.4. This is done by listing definitions for appropriate patterns. *Mathematica* automatically sorts them so that the specific ones are applied before the general. This allows the following recursive program for the factorial function:

In[141]:= **myFactorial[n_Integer?NonNegative] := n myFactorial[n - 1]**

In[142]:= **myFactorial[0] = 1;**

In[143]:= **myFactorial[10]**

Out[143]= 3628800

The command

In[144]:= **? myFactorial**

Global`myFactorial

myFactorial[0] = 1

myFactorial[(n_Integer)?NonNegative] := n*myFactorial[n - 1]

shows us the order in which the definitions are applied by *Mathematica*. We see that they are ordered such that the more specific definition is used before the more general one. This assures that the end condition is reached–and the algorithm terminates.

Large recursions can trigger an internal security limit.

In[145]:= **myFactorial[300]**

$RecursionLimit::reclim : Recursion depth of 256 exceeded.

Out[145]= 556203030714528117982157513075556570183842711336512883700395·
15243292095283955303008059152370574723282623082366303381970·
97829634401443145981610309370148732792447898675081845435004·
24469331618247148406583202183019314395508496150245243330615·
37654094389547013509047896956484542575251932909492919021399·
58043611711050489276742955535958345570224000114013025111650·
11731788110884032719420438135630831167687120808957899431632·
86505839081435386529923870798034282426460016036787055842956·
77731123611698200576000000000000000000000000000000000000000·
00000000000000000000000000 myFactorial[46]

If the recursion will terminate for sure, the limit can be enlarged (or even set to Infinity) by changing the global variable $RecursionLimit.

In[146]:= **$RecursionLimit = 10^3;**

In[147]:= **myFactorial[300]**

Out[147]= 306057512216440636035370461297268629388588804173576999416776·
74125947653317671686746551529142247757334993914788870172636·
88642639077590031542268429279069745598412254769302719546040·
08012215776252176854255965356903506788725264321896264299365·
20457644883038890975394348962543605322598077652127082243763·
94491201286786753683057122936819436499564604981664502277165·
00185176546469340112226034729724066333258583506870150169794·
16885035375213755491028912640715715483028228493795263658014·
52352331569364822334367992545940952768206080622328123873838·
80817049600000000000000000000000000000000000000000000000000·
00000000000000000000000000

Another example is the recursive calculation of Fibonacci numbers:

In[148]:= **fib1[0] = 0;**
fib1[1] = 1;
fib1[n_Integer?NonNegative] := fib1[n - 1] + fib1[n - 2]

In[151]:= **fib1[6]**

Out[151]= 8

But this implementation is useless because the time needed to calculate fib1 for larger arguments grows exponentially. This happens because new recursions begin on the right-hand side of the definition for each summand, thereby calculating the same values over and over again. The following implementation stores the already calculated values dynamically. This speeds up the calculation.

```
In[152]:=  fib2[0] = 0;
           fib2[1] = 1;
           fib2[n_Integer?NonNegative] :=
            fib2[n] = fib2[n - 1] + fib2[n - 2]

In[155]:=  fib2[6]

Out[155]=  8

In[156]:=  ?fib2

           Global`fib2

           fib2[0] = 0

           fib2[1] = 1

           fib2[2] = 1

           fib2[3] = 2

           fib2[4] = 3

           fib2[5] = 5

           fib2[6] = 8

           fib2[(n_Integer)?NonNegative] := fib2[n] = fib2[n - 1] + fib2[n - 2]
```

The efficiency of both variations is dramatically different. (For a fair comparison, values which have already been calculated for fib2 must first be cleared.)

```
In[157]:=  Clear[fib2];
           fib2[0] = 0;
           fib2[1] = 1;
           fib2[n_Integer?NonNegative] :=
            fib2[n] = fib2[n - 1] + fib2[n - 2]

In[161]:=  Timing[fib1[26]]

Out[161]=  {18.1167 Second, 121393}

In[162]:=  Timing[fib2[26]]

Out[162]=  {0. Second, 121393}
```

In the following section, we will discuss a procedural implementation which increases efficiency but reduces readability.

■ 4.4.4 Procedural Programming

Mathematica contains the branches If, Which, Switch and the loops Do, While, For for procedural programming. Do and While can be useful. The For loop is a concession to C programmers; it often leads to badly structured programs.

If can process two to four arguments: If[*condition*, *ifTrue*, *ifFalse*, *otherwise*].

```
In[163]:= Table[If[PrimeQ[n], n, FactorInteger[n]], {n, 2, 10}]
Out[163]= {2, 3, {{2, 2}}, 5, {{2, 1}, {3, 1}},
          7, {{2, 3}}, {{3, 2}}, {{2, 1}, {5, 1}}}
```

The variation *otherwise* handles cases where the test function does not evaluate to True or False:

```
In[164]:= If[NonNegative[#], "nonnegative", "negative", "unknown"] & /@
          {-1, 0, 1, a}
Out[164]= {negative, nonnegative, nonnegative, unknown}
```

Which processes an even number of arguments where every test is followed by the result that must be returned if the test yields True. The tests are processed from left to right until the first True.

Let us look at the function

```
In[165]:= intervals[x_] = Which[x < 0, 0, x < 1, 1, x < 2, 2];
```

and evaluate it for the elements of the following list:

```
In[166]:= intervals /@ {-.5, .5, 1.5, 2.5}
Out[166]= {0, 1, 2, Null}
```

Values ≥ 2 are not anticipated by this function: Which yields the symbol Null. We can catch such exceptions by entering True as the last test.

```
In[167]:= intervalTest[x_] =
          Which[x < 0, 0, x < 1, 1, x < 2, 2, True, "outside"];
```

In[168]:= **intervalTest /@ {-.5, .5, 1.5, 2.5}**

Out[168]= {0, 1, 2, outside}

Switch tests a given expression on patterns. After the expression, pairs of patterns and the corresponding results follow. Here the exceptions can be caught with a blank.

In[169]:= **analyze[x_] :=**
        Switch[x, _^2, "quadratic", _^3, "cubic", _, "other"]

In[170]:= **analyze /@ {a, a^2, a^3, a^6}**

Out[170]= {other, quadratic, cubic, other}

Do is analogous to Table, except that it does not yield a result. We illustrate such a loop with the Print function.

In[171]:= **Do[Print[1 / x], {x, 5}]**

1

$\frac{1}{2}$

$\frac{1}{3}$

$\frac{1}{4}$

$\frac{1}{5}$

The following little program calculates Fibonacci numbers with a Do loop, by beginning with the first two (0 and 1) and calculating the higher ones by adding the two preceding ones. By working with a list containing the values of two successive Fibonacci numbers, we arrive at a very elegant program. It is more efficient than the recursive ones regarding memory and compute time. On the other hand, with the recursive program, we can immediately see how it works, whereas here we have to think about it a bit first.

In[172]:= **fib3[n_] := Module[{fn1 = 0, fn2 = 1},**
        Do[{fn1, fn2} = {fn1 + fn2, fn1}, {n}];
        fn1]

In[173]:= **fib3[200]**

Out[173]= 280571172992510140037611932413038677189525

There are even faster methods for calculating Fibonacci numbers. The built-in function `Fibonacci` uses such an algorithm.

The `While` function uses a test as the first argument and as the second a compound expression (single expressions divided by semicolons). Here the various possibilities for manipulating iteration variables (see Section 2.4.4 of the *Mathematica Book*) can be useful, for example `++`.

In[174]:= **Module[{n = 1, t}, t = n; While[n <= 4, t = x + 1 / t; n++]; t]**

Out[174]= $x + \dfrac{1}{x + \dfrac{1}{x + \frac{1}{1+x}}}$

If necessary the program flow may be controlled using `Return`, `Continue`, `Break`, and `Catch`/`Throw`.

■ 4.4.5 Modularity

For the developer of a *Mathematica* package, the danger of name collisions does not only exist for auxiliary variables (which can be localized using `Module`), but also for the function names themselves. It might happen that two programmers of packages use the same name for functions which solve very different tasks. For this reason, *Mathematica* places every name into a so-called *context*. Each package creates its own contexts and uses these for its names.

Context names are marked with back quotes (`` ` ``) and organized hierarchically. If a `` ` `` comes first, it is to be taken relatively. Two contexts are predefined:
• `Global`` contains the names entered by the user during the working session,
• `System`` contains the names built into the kernel.

The function `Context` shows the context of a name:

In[175]:= **Context[x]**

Out[175]= Global`

In[176]:= **Context[Integrate]**

Out[176]= System`

We can manually introduce the name `x` into the context `myContext``. This makes it different from an `x` in the `Global`` context:

In[177]:= **myContext`x - x**

Out[177]= $-x + myContext`x$

A *Mathematica* package must use `BeginPackage-EndPackage` and `Begin-End` in such a way that the exported names are placed in the context of the package and the hidden names in a private sub-context. The following template shows how to do this:

```
BeginPackage["PackageName`", {"Needed1`", "Needed2`", …}]

Function1::usage = "Function1[x] calculates …"

…

Begin["`Private`"]

hiddenVariable = …

Function1[x_] := …

…

End[]

EndPackage[]
```

`PackageName`` stands for the context name of the package and should be chosen to reflect its contents. Conventionally, the corresponding package file (see below) should be named `PackageName.m`.

The list of context names `{"Needed1`", "Needed2`", …}` is only necessary if the package is based on other packages which must be loaded automatically. Otherwise it can be left out.

A short documentation of every exported object (`Function1`) must be placed between `BeginPackage` and `Begin`. These `usage` statements are defined as strings containing the documentations. They can be accessed by the user of the package (e.g. `?Function1`).

`Begin` opens a private sub-context which automatically hides new names introduced here (`hiddenVariable`). Because the context name of the package (`PackageName``) should be unique when the package is loaded, the relative sub-context name `` `Private` `` can always be used.

A simple package will usually be passed on in the form of a formatted notebook, which contains the *Mathematica* code and examples, and a package file. In order that everything functions properly, the code in the notebook, i.e. all the input cells between `BeginPack-age` and `EndPackage`, must be marked as initialization cells (menu **Cell > Cell Properties > Initialization Cell**). When saving, a message appears asking whether the initialization cells should be saved into a package file. After selecting **Create Auto Save Package**, *Mathematica* will create a file named `PackageName.m`, which can be loaded as usual with `<<PackageName`. Changes in the notebook file are automatically carried over to the package file. Both files should be placed into the `Applications` or `Autoload` directory (sub-directories of `AddOns` in *Mathematica*'s installation directory). If these directories are write protected, the personal *Mathematica* directory can be chosen. In this way the file will safely be found and in case of `Autoload` it will even be loaded automatically when a kernel is launched.

Names and arguments of exported functions should be chosen similar to existing *Mathematica* functions, so that the user will easily become acquainted with the new functions.

■ 4.4.6 Compiling Numerical Calculations

The efficiency of numerical calculations can be increased with the `Compile` function. Its arguments are analogous to `Function` (see the section about pure functions). Additional information about the types of arguments can be given.

The following calculation is speeded up through compilation by about factor 4. First, we compile the expression.

In[178]:= **compiledExpression = Compile$\left[\mathbf{x}, \dfrac{1 + \mathbf{x} + \mathbf{x}^2}{2 + \mathbf{x} - 5\,\mathbf{x}^2 - \mathbf{x}^3}\right]$**

Out[178]= $\text{CompiledFunction}\left[\{x\}, \dfrac{1 + x + x^2}{2 + x - 5\,x^2 - x^3}, -\text{CompiledCode}-\right]$

This object can be applied to an argument just like a pure function.

In[179]:= **compiledExpression[1.5]**

Out[179]= -0.426966

For a comparison with an uncompiled variation, we loop a couple of times.

In[180]:= **Timing[Do[compiledExpression[1.5], {10000}]]**

Out[180]= $\{0.3\ \text{Second}, \text{Null}\}$

In[181]:= $\text{Timing}\Big[\text{Do}\Big[\dfrac{1 + 1.5 + 1.5^2}{2 + 1.5 - 5\,1.5^2 - 1.5^3},\ \{10000\}\Big]\Big]$

Out[181]= $\{1.25\ \text{Second},\ \text{Null}\}$

For complex arguments, the compilation would look like this:

In[182]:= $\text{compiledComplexExpression} =$
$\text{Compile}\Big[\{\{x,\ \_\text{Complex}\}\},\ \dfrac{1 + x + x^2}{2 + x - 5\,x^2 - x^3}\Big]$

Out[182]= $\text{CompiledFunction}\Big[\{x\},\ \dfrac{1 + x + x^2}{2 + x - 5\,x^2 - x^3},\ -\text{CompiledCode}-\Big]$

In[183]:= $\text{compiledComplexExpression}[2. + 3.\ \text{I}]$

Out[183]= $-0.114217 + 0.099489\ \text{I}$

■ Exercises

● Newton's Algorithm

The application of Newton's algorithm to the polynomial $x^2 - 3$ yields an approximation for $\sqrt{3}$. Program it first functionally, then procedurally.

● Fibonacci Numbers

Program the calculation of the Fibonacci numbers using a procedural algorithm which does not work with lists (like `fib3`).

Compare the timings of all program variations. Do not forget that the recursive implementation `fib2` stores all calculated values. They must therefore be deleted before a comparison is made.

● Packages

Create a package which defines and exports the function for gradient calculation (exercises to Section 3.2). Use the name `Grad` (`Gradient` is already taken by an option of `FindMinimum`).

● Programming

If you have a small programming exercise for a procedural language handy, try to solve it in *Mathematica*. Consider if a functional or a rule-based algorithm would also be possible.

■ 4.5 Further Information

■ 4.5.1 Internet

The Web site of Wolfram Research (http://www.wolfram.com/), the company behind *Mathematica*, is worth a visit. There you will find among other things up-to-date information about the program and FAQs (frequently asked questions). It is a good idea to consult the FAQ page before contacting the support team support@wolfram.com (you should indicate your license number $LicenseID, version $Version, and operating system).

The Usenet conference comp.soft-sys.math.mathematica is frequented by beginners and experts to *Mathematica*.

■ 4.5.2 MathSource

Wolfram Research's *MathSource* server at http://www.mathsource.com/ contains notebooks and packages for various types of applications. Many of them are free.

■ 4.5.3 Literature

The amount of literature about *Mathematica* is growing quickly, with well over one hundred books extant at present. You will find an up-to-date list on the Web site of Wolfram Research. Go to http://www.wolfram.com/ and follow the links > Products & Store > *Mathematica* Bookstore.

You might also be interested in the *Mathematica Journal* which is published on the Web. Check out the site http://www.mathematica-journal.com/.

■ Index

Mathematica-objects are printed in **bold Courier font**, file names, packages and commands in `plain Courier font`, menu commands and elements of the *Help Browser* in **bold Times font**.